Daniel Sloughter

Neoplatonism and the Art of Calculation in Medieval Iceland

Research in Medieval and Early Modern Culture XLI

Studies in Medieval and Early Modern Culture XCI

Daniel Sloughter

Neoplatonism and the Art of Calculation in Medieval Iceland

Algorithms, Geometry, and the Platonic Elements in the Medieval North

DE GRUYTER

MIP
University Press | Kalamazoo
MEDIEVAL INSTITUTE PUBLICATIONS
Western Michigan University

ISBN 978-1-5015-2408-0
ISBN 978-1-5015-2237-6 (PDF)
DOI https://doi.org/10.1515/9781501522376

Library of Congress Control Number: 2026944814

Bibliographic information published by the Deutsche Nationalbibliothek
The Deutsche Nationalbibliothek lists this publication in the Deutsche Nationalbibliografie; detailed bibliographic data are available on the internet at http://dnb.dnb.de.

De Gruyter and Walter de Gruyter GmbH are part of De Gruyter Brill.
www.degruyterbrill.com

Questions about General Product Safety Regulation:
productsafety@degruyterbrill.com

Cover illustration: Isidor, De natura rerum, Staatsbibliothek Bamberg
Msc.Nat.1, f. 15r, photo: Gerald Raab (Public Domain Mark 1.0)
Printing: CPI books GmbH, Leck

MIX
Papier | Fördert
gute Waldnutzung
FSC® C083411
FSC
www.fsc.org

To Merle Farrier (1947–2025), a teacher without peer.

Preface

The small island of Viðey sits in the bay just a short ferry ride from Reykjavík. Uninhabited these days, with only a few structures still standing, it is best known now as the location of the Imagine Peace Tower, which, at certain times of the year, shines a beam of light into the sky in remembrance of John Lennon. But, from its founding in 1226 until it was attacked and desecrated by agents of the Danish king on Whitsunday in 1539, Viðey was the site of Viðeyjarklaustur, an Augustinian monastery and a center for learning in medieval Iceland.[1]

The Old Norse manuscript GKS 1812 4to has long been associated with Viðeyjarklaustur. The manuscript, or at least its component parts, may have been passed between family members originally connected with the monastery until Bishop Brynjólfur Sveinsson of Skálholt sent it to Copenhagen in 1662. Or it may have been amongst the documents taken from Viðey during the Danish ransacking in 1539, finding its way first to the royal manor at Bessastaðir and later to the possession of Bishop Brynjólfur. However, although much has been made of circumstantial evidence of links between the manuscript and Viðey, there is little that can be said about its origins with much certainty.[2] Moreover, the manuscript as we have it today is a composite of three sections, composed in different centuries, which were bound together at some unknown time after the composition of the last section in the early 14th century and before the entire manuscript was sent to Copenhagen in 1662.[3] Indeed, it may not have even had its current form before the Danish attack on Viðeyjarklaustur. At the same time, the contents of the volume point to its intended use as a collection of texts useful for education in various sciences of the day. Hence, even if the manuscript, or its components, never resided in Viðey, it is nevertheless conceptually convenient to regard it as a monastic manuscript composed for educational purposes.

The goal of this volume is to examine a number of considerations, both historical and philosophical, which arise from two mathematical tracts found in GKS 1812 4to. The first of these two texts is a presentation of the methods of the Indo-Arabic

---

1 Steinunn Kristjánsdóttir, "The Dissolution of the Monastic Houses in Iceland," *Religions* 15, no. 7:771 (2024), accessed December 10, 2025, https://doi.org/10.3390/rel15070771.

2 For a detailed discussion of the provenance of GKS 1812, see: Gunnar Harðarson, "Medieval Encyclopedic Literature and Icelandic Manuscripts," in *A World in Fragments: Studies on the Encyclopedic Manuscript GKS 1812 4to*, ed. Gunnar Harðarson et al. (Reykjavík: Stofnun Árna Magnússonar í Íslenskum Fræðum, 2021), 13–38.

3 Svanhildur Óskarsdóttir and Vasareé Rastonis, "Fragments United: The Codicology of GKS 1812 4to," in *A World in Fragments: Studies on the Encyclopedic Manuscript GKS 1812 4to*, ed. Gunnar Harðarson et al. (Reykjavík: Stofnun Árna Magnússonar í Íslenskum Fræðum, 2021), 61.

 | https://doi.org/10.1515/9781501522376-202

Fig. 1: Viðey as seen from Reykjavík (author's photograph)

system of numeration, and the second is a discussion of some computations concerning the relationships of certain inscribed and circumscribed circles and equilateral triangles.

To those ends, I provide transcriptions, each with critical apparatus, of the two texts. These transcriptions fall roughly into the category that the *Medieval Nordic Text Archive* (*Menota*) calls the *diplomatic level.*[4] This is a transcription level which sits between a straight reproduction of the abbreviations and characters used in the original manuscript (the *facsimile level*) and an attempt to render the text in some standardized version of medieval Icelandic (the *normalized level*). This is not an attempt to create a critical edition of either text. My interest here is in presenting a version of these texts as they appeared in a particular monastic setting. The critical apparati provide alternative readings from the small number of other extant manuscripts which also contain the two texts. A diplomatic transcript of the versions of these texts as they appear in GKS 1812 is also of historical interest in underscoring the distinctiveness of the text and in illustrating the variance in 14th-century Icelandic scripts. A normalized version of the *Algorismus* would differ little from the one Kristín Bjarnadóttir has published in modern Icelandic orthography, based on

4 Haraldur Bernharðsson and Odd Einar Haugen, "Levels of Text Representation," chap. 4, 3rd ed., ed. Odd Einar Haugen (Bergen, Norway: Medieval Nordic Text Archive, 2019), accessed December 10, 2025, https://menota.org/HB3_ch4.xml.

a version found in the manuscript AM 544,[5] and the critical apparatus for such a version would not reveal as much about the Norwegian influence on the version in GKS 1812, nor of the variations in the representation of numbers between the manuscripts.

The main text considered here, a version of the Old Norse *Algorismus*, is, for the most part, a translation of a version of the *Carmen de algorismo* of Alexander of Villedieu. For this reason, I have included a transcript of one version of the latter treatise. There exist a large number of variant versions of the *Carmen de algorismo*. What I provide here is but a transcription of one version, with a critical apparatus restricted to a small set of well-known variants.

Neither of the Old Norse texts considered here is, in itself, of particular interest in the broader story of the advancement of mathematical or scientific reasoning in the Middle Ages. However, they are of historical and philosophical interest in at least two regards. First, both texts are illustrative of the spread of practical algorithmic thinking to the edges of medieval Europe. This is particularly true of the *Algorismus* and, as such, provides a starting point for discussing the relationship between speculative and practical mathematics (or algorithmics). Second, the final section of the *Algorismus* contains a neoplatonic description of the creation of the world and, in particular, the four elements (that is, earth, water, air, and fire). Such a passage has not been found in any other version of similar medieval texts, and so leads to questions of its provenance and its place in an ostensibly mathematical work. Moreover, it raises questions about where to locate the dividing line between mathematics and numerology.

If not noted otherwise, translations in the following pages are mine. I use translations from other sources when suitable ones are available, even if that sometimes results in inconsistencies in the translation of some terms. On a few occasions, however, I use my own translations even though the passages in question have been translated elsewhere. In particular, I use my own translations for some verse quotations when I want to emphasize the literal meaning, whereas other translators may need to take into consideration additional metrical and contextual considerations.

I have not capitalized 'platonism,' 'neoplatonism,' and 'pythagorean.' In the context of the discussions in this book, these terms refer not so much to specific philosophical schools as they do to more general philosophical points of view. As such, I

5 Kristín Bjarnadóttir, "Algorismus: Fornt stærðfræðirit í íslenskum handritum," *NETLA*, 2004, accessed December 10, 2025, http://wayback.vefsafn.is/wayback/20201017180549/https://netla.hi.is/greinar/2004/001/index.htm.

avoid attempting to make finer distinctions between the various schools and phases of these approaches to philosophy.

Earlier versions of the transcript (without the extensive critical apparatus) and translation of the *Algorismus*, as well as some other parts of the present text, appeared in my master's thesis at the University of Iceland[6] and in an article published in *Gripla*.[7]

6 Daniel Sloughter, "Algorismus in GKS 1812" (Master's Thesis, Háskóli Íslands, 2020), accessed December 9, 2025, https://skemman.is/handle/1946/37284.

7 Daniel Sloughter, "*Algorismus* in GKS 1812 4to," *Gripla* XXXII (2021): 73–100.

Acknowledgements

I am indebted to many individuals who have guided me in the research that has led to this work. At the University of Iceland and the Árni Magnússon Institute for Icelandic Studies, Haraldur Bernharðsson first introduced me to the Old Norse *Algorismus* in GKS 1812, Gunnar Harðarson provided invaluable advice and encouragement as my master's thesis advisor, Þórdis Edda Jóhannesdóttir introduced me to the reading of medieval Icelandic manuscripts and answered numerous questions as I worked through the text of the *Algorismus*, Margaret Cormack gave me inspiration to keep working and provided many helpful comments on translating Old Icelandic into modern English, Karl Óskar Ólafsson provided me with access to the original manuscript for a final check of my transcription, and Sigurður Jónsson took new color photographs of the manuscript to aid in that final pass through the text.

I owe a special thanks to the many individuals at De Gruyter Brill who provided me with suggestions, guidance, encouragement, criticisms, and technical assistance. Among them are Robert Forke, Christine Henschel, Teodor Borsa, and Isabell Väth. Two anonymous reviewers provided invaluable assistance in identifying needed corrections, clarifications, and additions. Their input has greatly improved the text. In addition, Ilse Schweitzer's thorough review of the text not only discovered errors in the text, but also provided numerous suggestions that have enhanced the coherence and clarity of the book. Theresa Whitaker, editor-in-chief of Medieval Institute Publications, assisted in enhancing several of the images in the text, and also suggested a number of additional corrections.

Any faults which remain are completely my own.

Contents

List of Figures

1 Background

1.1 GKS 1812

In the late 11th century, Adam of Bremen described Iceland in very bleak terms. From his perspective in the northern reaches of the European continent, Iceland was a remote island in the far north inhabited by people who lived in poverty and holy simplicity:

> Est autem insula permaxima, ita ut populos infra se multos contineat, qui solo pecorum fetu vivunt eorumque vellere teguntur; nullae ibi fruges, minima lignorum copia. Propterea in subterraneis habitant speluncis, communi tecto [et victu] et strato gaudentes cum pecoribus suis.[1]
>
> [This island, however, is so very large that it has on it many peoples, who make a living only by raising cattle and who clothe themselves with their pelts. No crops are grown there; the supply of wood is very meager. On this account the people dwell in underground caves, glad to have roof and food and bed in common with their cattle.[2]]

The *subterraneae speluncae* to which Adam refers were, in fact, likely the sod homes which the poor Icelandic crofters shared with their livestock. Yet even before Adam wrote his history of the bishops of Hamburg-Bremen, a first attempt had been made in 1030 by the archbishop of Hamburg-Bremen to establish a monastery in Iceland.[3] Although this particular monastery, Bæjarklaustur, would be short-lived (lasting only nineteen years), thirteen more monasteries would follow.[4] Not all of these were successful, but among those was Viðeyjarklaustur, an Augustinian monastery situated on the small island of Viðey a short distance from Reykjavík. Established in 1226 with the aid of some of the most well-known chieftains of medieval Iceland, the monastery of Viðey survived until the Reformation.[5]

When thinking of the culture existing in medieval Iceland, what typically comes to mind are either the myths concerning the Norse gods or the epic sagas which tell the legends of men such as Egill Skallagrímsson, Njáll Þorgeirsson, and Grettir Ásmundarson. However, the network of Icelandic monasteries functioned much as

1 Adam von Bremen, *Hamburgische Kirchengeschichte*, ed. Bernhard Schmeidler (Hanover and Leipzig: Hahnsche Buchhandlung, 1917), 272.

2 Adam of Bremen, *History of the Archbishops of Hamburg-Bremen*, trans. Francis J. Tschan (New York: Columbia University Press, 2002), 217.

3 Steinunn Kristjánsdóttir, *Monastic Iceland* (London and New York: Routledge, 2023), 34.

4 Steinunn Kristjánsdóttir, *Monastic Iceland*, 2.

5 Steinunn Kristjánsdóttir, *Monastic Iceland*, 81.

https://doi.org/10.1515/9781501522376-001

they did in other parts of Europe. Specifically, in addition to religious activities and managing farms, the monasteries served as centers for education. And, as in the rest of medieval Europe, education in Iceland centered on the arts of the trivium (grammar, logic, and rhetoric) and the quadrivium (arithmetic, geometry, music, and astronomy).[6] As elsewhere in Europe, such instruction required the copying of manuscripts compiling the accepted state of knowledge of these subjects.

One such manuscript, cataloged as *Gammel kongelig samling 1812 quarto* (GKS 1812 4to) after the bishop of Skálholt gave it to the Danish king in the 17th century, has long been associated with Viðeyjarklaustur. A composite manuscript composed of a number of texts produced during the 12th through 14th centuries, it is primarily the work of four scribes, whose distinct hands form the basis for dividing the work into four parts. Although its origins remain obscure, with even the date at which its three components were bound into a single text remaining a matter of conjecture, the nature of the texts included clearly indicates its educational purpose. Today the manuscript resides at the Árni Magnússon Institute for Icelandic Studies at the University of Iceland, having been returned to Iceland from Denmark in 1984.[7]

The manuscript falls broadly into the category of medieval encyclopedic literature.[8] Unlike our modern conception of an encyclopedia—which typically implies an alphabetic compilation of the current state of knowledge on a broad swath of topics, either providing an overview of knowledge in general or from a particular discipline—the typical medieval encyclopedia consisted of texts useful for a liberal education in the trivium or quadrivium.[9] More particularly, the term describes medieval Icelandic manuscripts comprising texts from the quadrivium, natural history, or medicine.[10] The oldest part of the GKS 1812, referred to as Section IV, consists of eleven folios (24r to 34v).[11] The bulk of this section contains *Rímbegla*, a treatise devoted primarily to calculations involving calendars.[12] The next oldest part, Section III, dating from the second quarter of the 13th century, consists of only four

6 Steinunn Kristjánsdóttir, *Monastic Iceland*, 181.

7 For background on the manuscript, see Gunnar Harðarson, "Medieval Encyclopedic Literature"; for details on the scribes, see Haraldur Bernharðsson, "Scribes and Scribal Practice in GKS1812 4to," in *A World in Fragments: Studies on the Encyclopedic Manuscript GKS 1812 4to*, ed. Gunnar Harðarson et al. (Reykjavík: Stofnun Árna Magnússonar í Íslenskum Fræðum, 2021), 61–113. The preface in the same volume contains further background information.

8 Margaret Clunies Ross and Rudolf Simek, "Encyclopedic Literature," in *Medieval Scandinavia: An Encyclopedia (Encyclopedias of the Middle Ages)*, ed. Phillip Pulsiano and Kirsten Wolf (Oxon: Routledge, 2016), 164–66.

9 Ross and Simek, "Encyclopedic Literature," 164.

10 Gunnar Harðarson, "Medieval Encyclopedic Literature," 13.

11 Haraldur Bernharðsson, "Scribes and Scribal Practice," 63.

12 Haraldur Bernharðsson, "Scribes and Scribal Practice," 64.

folios (5r to 6v and 35r to 36v).[13] The text contains more computational material, along with calendars, a world map, and a list of certain Icelandic priests.[14] Sections I (folios 1r to 4v, 7r to 13r, 18v1 to 19r13, and 22r10 to 23v36) and II (folios 13v to 18r, 19r13 to 22r9, and 1vb23–40) date from the 14th century, possibly from the latter part of the first half of that century, and were written by two contemporaneous scribes who collaborated on some of their work.[15] These sections contain more computational and calendrical material; astronomical works, such as the *Stjǫrnu-Odda tal* and diagrams of the planets and the Zodiac; and a diagram of philosophy.[16] In addition, these sections contain two mathematical texts: the *Algorismus*, a primer on numeration and arithmetic using the Indo-Arabic system of notation, in Section II (folios 13v to 16v) and, immediately before, a diagram and discussion of inscribed geometrical figures in Section I (folio 13r).

1.2 The *Algorismus*: From India to Iceland

The story of the Old Norse *Algorismus* found in GKS 1812 starts some 500 years earlier with Muḥammed ibn Mūsā al-Khwārizmī. Little is known of his personal life. We know that he lived during the first half of the 9th century and worked in the Bayt al-Ḥikma ("House of Wisdom") in Baghdad.[17] His toponym, al-Khwārizmī, indicates that he was born in what is now the city of Khiva in Uzbekistan, known at that time as Khwarezm, a merchant town along the ancient Silk Road trade route. Although few details are known of his life, his scientific works are well known, even to those who have never heard his name. A polymath, he produced important works in mathematics, astronomy, and geography. In the Middle Ages, Western Europe knew him for two mathematical works: *Kitāb al-Jabr wa-al-muqābala*[18] and *De numero Indorum.*[19] The first is the starting point of the modern mathematics of algebra; the second, now lost in its original Arabic form, was the source of numerous treatises in the West on the uses of the Indo-Arabic numerals. As but one witness to the influ-

13 Haraldur Bernharðsson, "Scribes and Scribal Practice," 63.

14 Haraldur Bernharðsson, "Scribes and Scribal Practice," 91.

15 Haraldur Bernharðsson, "Scribes and Scribal Practice," 63, 113.

16 Haraldur Bernharðsson, "Scribes and Scribal Practice," 102–3, 108; On the diagram of philosophy, see Gunnar Harðarson, "A *Divisio Philosophiae* in the Medieval Icelandic Manuscript GKS 1812°," *Cahiers de l'Institut du Moyen-Âge Grec et Latin*, no. 84 (2015): 2–21.

17 J. Vernet, "al-KHwārazmī," in *The Enclyclopaedia of Islam*, ed. C. E. Bosworth et al., vol. 4 (Leiden: E. J. Brill, 1997), 1070–71.

18 Roshdi Rashed, "Algebra," chap. 11 in *Encyclopedia of the History of Arabic Science*, ed. Roshdi Rashed, vol. 2 (London: Taylor & Francis, 1996), 345.

19 The original title in Arabic is uncertain. For two possibilities, see Vernet, "al-KHwārazmī," 1070.

ence of these works, one need only note that the modern term 'algebra' is derived from the title of the first and the term 'algorismus,' the source of the modern term 'algorithm,' became associated with the Latin versions of the second, and is most likely derived from the toponym of Muḥammed ibn Mūsā.[20]

The method by which the system of writing and performing arithmetical operations with the Indo-Arabic numerals, what became known in medieval Western Europe as the Indian calculus, spread from India to the rest of the world is a complicated story. As David Smith and Louis Karpinski have noted, in all probability, "it was the trader rather than the scholar who carried these numerals from their original habitat to various commercial centers."[21] For example, Avicenna (Ibn Sina, ca. 970–1037) relates that, when he was young, his father sent him to a local merchant in Bukhara, Uzbekistan,[22] to learn the Indian calculus.[23] As for the scientific world, it is likely that the numerals were passed on to Arab scholars, in the form of Indian astronomical tables, sometime in the 8th century[24] and were later brought to Spain sometime in the second half of the 10th century.[25] Leonardo Pisano (1170–1240), known since the 19th century as Fibonacci,[26] learned the art of the Indian numbers from Arab scientists in northern Africa and wrote the most famous book on the topic, the *Liber abaci*, in 1202.[27] However, Leonardo's treatment was not disseminated widely at first,[28] and it would be from Spain, in the form of translations of the *De numero Indorum* of al-Khwārizmī, that most of Europe would learn how to use the new system.

The earliest versions of the Indian calculus in Western Europe may have been close to the original work of al-Khwārizmī. But, even in their earliest form, they

20 Joseph Toussaint Reinaud, "Mémoire géographique, historique et scientifique sur l'Inde, antéreieurment au milieu du XIe siécle de l'ére chrétienne, d'aprés les écrivains arabes, persans et chinois," in *Mémoires de l'Institute National de France*, vol. 18 (Paris: Académie des Inscriptions et Belles-Lettres, 1849), 304–5.

21 David Eugene Smith and Louis Charles Karpinski, *The Hindu Arabic Numerals* (Boston, MA: Ginn & Company, 1911), 99.

22 Like Khwarezm, a merchant town on the Silk Road.

23 M. F. Woepcke, *Sur l'introduction de l'artithmétique Indienne en Occident et sur deux documents importants* (Rome: Imprimerie des Sciences Mathématiques et Physiques, 1859), 52.

24 Smith and Karpinski, *The Hindu Arabic Numerals*, 92.

25 Leonardo Pisano, *Fibonacci's Liber Abaci*, trans. L. E. Sigler (New York: Springer, 2002), 3.

26 Smith and Karpinski, *The Hindu Arabic Numerals*, 130.

27 Leonardo Pisano, *Fibonacci's Liber Abaci*, 3-4, 15-16. What we know of Leonardo's education in the Indian calculus comes from a single paragraph in the *Liber abaci*. Although this passage is vague, and the manuscripts not without corruptions, it is clear that he learned the techniques of the new calculus from study and the "give-and-take of of disputation." See: Richard E. Grimm, "The Autobiography of Leonardo Pisano," *The Fibonacci Quarterly* 11, no. 1 (1973): 99–104.

28 Smith and Karpinski, *The Hindu Arabic Numerals*, 131.

were never direct translations. Over the next three centuries or so, subsequent versions would adapt the original into a variety of forms. Some of these added material, while others subtracted; some provided additional explanations and examples, while others were more terse; and some were even turned into verse.

In her doctoral thesis in 1914, Suzan Rose Benedict identified thirty-eight different medieval treatises on the Indian calculus, dating from the 7th century to the 15th century.[29] The majority of these are in Latin, although she begins her catalogue with an astronomical text in Hindu and includes works in Arabic, Hebrew, French, Old Norse, and English as well. More have been discovered since 1914.[30] The four oldest Latin treatises date from the first half of the 12th century: the *Dixit Algorizmi*, the *Liber ysagogarum Alchorismi*, the *Liber Alchorismi*, and the *Liber pulueris*.[31] Michel Chasles was the first modern scholar to realize that the *Liber abaci* of Leonardo Pisano was not the first text on Indian calculus introduced to Europe,[32] writing in 1837 that the *Liber Alchorismi* was translated in the first half of the 12th century by John of Seville, a well-known Latin translator.[33] This attribution was accepted for many years, but most likely stems from an assumption made by a later scribe, and now *Liber Alchorismi* is regarded as the work of John of Toledo.[34]

Most consider the *Dixit Algorizmi* to be the oldest of these four Latin treatises.[35] It is often referred to as the *De numero Indorum*, which seems to identify the work directly with the original work of al-Khwārizmī. In fact, it is a hybrid text, drawing upon texts of Boethius and Nicomachus of Gerasa as well as that of al-Khwārizmī.[36] This treatise begins with a lengthy explanation of the representation of positive integer numbers (that is, numeration) with the Indo-Arabic numerals. The final example shows the simplicity of the system, interpreting the number 1,180,703,051,492,863, created from just the ten basic characters 0, 1, 2, 3, 4, 5, 6, 7, 8, and 9.[37] From here,

29 Suzan Rose Benedict, *A Comparative Study of the Early Treatises Introducing into Europe the Hindu Art of Reckoning* (Concord, New Hampshire: The Rumford Press, 1914).

30 See, for examples, Louis C. Karpinski, "Two Twelfth Century Algorisms," *Isis* 3, no. 3 (1921): 396–413; Louis C. Karpinski, "A Thirteenth Century Algorism in French Verse," *Isis* 11, no. 1 (1928): 45–84.

31 André Allard, ed., *Muhammad Ibn Mūsā Al-Khwārizmī: Le Calcul Indien (Algorismus)* (Namur, Belgium: Société des Études Classiques, 1992), II–III.

32 Allard, *Le Calcul Indien (Algorismus)*, VIII.

33 Michel Chasles, *Aperçu historique sur l'origine et le développement des méthodes en géometrie* (Brussels: M. Hayes, Imprimeur de l'Académie Royale, 1837), 510.

34 André Allard, "The Influence of Arabic Mathematics in the Medieval West," chap. 16 in *Encyclopedia of the History of Arabic Science*, ed. Roshdi Rashed, vol. 2 (London: Taylor & Francis, 1996), 536.

35 Allard, *Le Calcul Indien (Algorismus)*, III.

36 Allard, "The Influence of Arabic Mathematics," 516.

37 Allard, *Le Calcul Indien (Algorismus)*, 6.

the treatise provides the algorithms for addition, subtraction, mediation (that is, dividing a number in half), duplation (that is, finding the double of a given number), multiplication, and division of positive integers, as well as multiplication and division of certain fractions. The statement, "nunc incipiemus tractare de multiplicatione fractionum, et earum divisione, et de extractione radicum, si Deus voluerit"[38] ["now we shall begin to treat the multiplication of fractions and their division and the extraction of roots, if God so wills"[39]] indicates that, most likely, the original Arabic text also treated algorithms for finding roots. However, the *Dixit Algorizmi* ends abruptly without considering these methods.

The *Liber ysagogarum Alchorismi* constitutes the first part of a treatise on astronomy, and is the shortest of these four works. It treats the same arithmetic operations as the *Dixit Algorizmi*, as well as finding square roots, but does not go into detail on the numeration system itself. The *Liber Alchorismi* is a much longer text. It includes all of the topics covered in the *Dixit Algorizmi*, along with a lengthier discussion of fractions and a section on finding square roots. The *Liber pulueris* is closely related to the *Liber Alchorismi*, and is most likely an older version derived from the same Latin sources.[40]

There are at least three more works which may be from the 12th century as well: a treatise belonging to the Salem Abbey in Germany[41] and two closely-related works found in the British Library.[42] The Salem treatise, after a short introduction to the system of numeration, explicitly states that there are seven operations, referred to as *species*, of arithmetic—that is, addition, subtraction, duplation, mediation, multiplication, division, and extraction of roots.[43] The work ends with a description of the relationship between these seven *species* and the seven gifts of the Holy Spirit.[44] The other two works mention only the first six of these operations, along with a brief description of numeration.

Four influential versions of the Indian calculus appeared in the 13th century: the *Liber abaci* of Leonardo Pisano; the *Carmen de algorismo* of Alexander of

38 Allard, *Le Calcul Indien (Algorismus)*, 16.

39 John N. Crossley and Alan S. Henry, "Thus Spake al-Khwārizmī: A Translation of the Text of Cambridge University Library Ms. Ii.vi.5," *Historia Mathematica* 17 (1990): 120.

40 Allard, "The Influence of Arabic Mathematics," 517.

41 Moritz Cantor, "Über einen Codex des Klosters Salem," *Zeitschrift für Mathematik und Physik*, no. 10 (1865): 1–16.

42 Karpinski, "Two Twelfth Century Algorisms."

43 Cantor, "Über einen Codex des Klosters Salem," 3.

44 Cantor, "Über einen Codex des Klosters Salem," 11.

Villedieu;[45] the *Tractatus de arte numerandi*, or *Algorismus vulgaris*, of John of Sacrobosco;[46] and *In algorismum vulgarem Johannis de Sacrobosco commentarius*, or *Expositio super algorismum*, of Peter of Dacia.[47]

The *Liber abaci* is the earliest of these, appearing first in 1202,[48] and is original to Leonardo. Although in many ways mathematically superior to the works derived from the treatise of al-Khwārizmī, it would not be widely appreciated until centuries later. Benedict speculates that this may have been because the mathematics was too advanced for the scholars of Western Europe at the time, or perhaps simply "because the monks, copying from monastery to monastery spread the other type of algorism."[49]

The existence of a large number of manuscripts containing copies of the *Carmen de algorismo* and the *Algorismus vulgaris* attests to their role in spreading the new calculation procedures widely.[50] Their respective authors, Alexander of Villedieu and John of Sacrobosco, both taught in Paris in the first half of the 13th century.[51] Both texts were written during this period, with the *Carmen de algorismo* clearly written earlier, as it is quoted in the *Algorismus vulgaris*.[52] Alexander of Villedieu was already well known for his book on grammar, the *Doctrinale puerorum*, and, like that text, the *Carmen de algorismo* was written in hexameter.[53] The short text, running to around 300 verses, discusses numeration only briefly and then proceeds to the same seven operations of arithmetic, in the same order, as the Salem treatise, although here they are referred to as *partes* rather than *species*.[54] Moreover, the extraction of roots is divided into square and cube roots.

45 Alexander de Villedieu, "Carmen de algorismo," in *Rara Mathematica*, ed. James Orchard Halliwell (London: John William Parker, 1839), 73–83; Robert Steele, ed., "Carmen de algorismo," in *The Earliest Arithmetics in English* (London: Oxford University Press, 1922), 72–80.

46 Iohannis de Sacro Bosco, "Tractatus de arte numerandi," in *Rara Mathematica*, ed. James Orchard Halliwell (London: John William Parker, 1839), 1–26; Iohannis de Sacro Bosco, "Algorismus," in *Petri Philomenae de Dacia et Petri de S. Audomaro: Opera quadrivialia*, ed. Fridericus Saaby Pedersen, vol. 1 (Copenhagen: Societas Linguae & Litterarum Danicarum, 1983), 165–201.

47 Petrus de Dacia, "Expositio super Algorismum," in *Petri Philomenae de Dacia et Petri de S. Audomaro: Opera quadrivialia*, ed. Fridericus Saaby Pedersen, vol. 1 (Copenhagen: Societas Linguae & Litterarum Danicarum, 1983), 69–163.

48 Leonardo Pisano, *Fibonacci's Liber Abaci*, 4.

49 Benedict, *A Comparative Study of the Early Treatises*, 120.

50 Allard, "The Influence of Arabic Mathematics," 523–24.

51 Benedict, *A Comparative Study of the Early Treatises*, 12–13.

52 Iohannis de Sacro Bosco, "Algorismus," 183; Alexander de Villedieu, "Carmen de algorismo," 74–75.

53 Benedict, *A Comparative Study of the Early Treatises*, 122.

54 Alexander de Villedieu, "Carmen de algorismo," 74.

The *Algorismus vulgaris* lists nine operations, referred to as *species*, of arithmetic, adding numeration and progression (that is, summing an arithmetic progression of numbers starting from 1 or 2).[55] The latter operation appears in a single manuscript of the *Liber Alchorismi*,[56] as well as in at least one manuscript of the *Carmen de algorismo*.[57] The *Algorismus vulgaris* explains the operations in greater detail and clarity than the *Carmen de algorismo*, yet, other than the section on progressions, provides few examples. The works may have been considered complementary, the *Carmen de algorismo* providing a brief, versified (and hence easier to memorize) form, while the *Algorismus vulgaris* supplied the clearer, more detailed, exposition.[58] For examples, the manuscript Ashmole 399 in the Bodleian Library at Oxford contains the *Carmen de algorismo* in folios 64r to 71r, with the *Algorismus vulgaris* in the margins of folios 64r to 66v;[59] and the manuscript Bodleian 177 contains parts of the *Algorismus vulgaris*, with each chapter followed by the corresponding passage from the *Carmen de algorismo*.[60]

In *Expositio super algorismum*, Peter of Dacia provides the examples missing in the *Carmen de algorismo* and *Algorismus vulgaris*, as well as more detailed explanations and background. Written in 1291[61] as a commentary on the work of John of Sacrobosco, the text treats the same operations as the *Algorismus vulgaris*, sometimes, as in the extraction of cube roots, with modifications.[62] Moreover, the examples in the text greatly enhance the often complex verbal explanations of the original. For example, in his English translation of the *Algorismus vulgaris*, Edward Grant comments, in regard to the description of finding cube roots, that "[t]his complicated

55 Iohannis de Sacro Bosco, "Algorismus," 174–75.

56 Baldassarre Boncompagni, ed., "Ionnis Hispalensis: Liber Algorismi de pratica arismetrice," in *Trattati d'aritmetica*, vol. II (Rome: Tipografia Delle Scienze Fisiche e Matematiche, 1857), 93–97.

57 For example, in the manuscript MS Latin 7475 in the Bibliothèque nationale de France, transcribed below in chapter 6. See: Alexander de Villa Dei, "Carmen de algorismo," in BNF Latin 7475 (Bibliothèque nationale de France), accessed December 10, 2025, https://archivesetmanuscrits.bnf.fr/ark:/12148/cc94612b.

58 Nadia Ambrosetti, "Algorithmic in the 12th Century: The Carmen de Algorismo by Alexander de Villa Dei," in *History and Philosophy of Computing: Third International Conference, 2015*, ed. Fabio Gadduci and Mirko Tavosanis (Cham, Switzerland: Springer, 2016), 81.

59 "MS. Ashmole 399, 64r–71r," in *Digital Bodleian* (Bodleian Library, University of Oxford), accessed December 10, 2025, https://digital.bodleian.ox.ac.uk/inquire/p/15a20263-f647-4913-9089-a9efbb37225c.

60 "MS. Bodleian 177, 45," in *Medieval Manuscripts in Oxford Libraries* (Bodleian Library, University of Oxford), accessed December 10, 2025, https://medieval.bodleian.ox.ac.uk/catalog/manuscript_1234.

61 Benedict, *A Comparative Study of the Early Treatises*, 17.

62 Bo Göran Johansson, "Cube Root Extraction in Medieval Mathematics," *Historia Mathematica* 38 (2011): 362.

verbal description is virtually unintelligible without a concrete example," and then provides one taken from the commentary of Peter of Dacia.[63]

In addition to the Latin texts, the 13th century saw the emergence of versions of the Indian calculus in Western European vernacular languages. One of the earliest of such vernacular versions is a 13th-century text in French verse.[64] It does not appear to reflect any other known treatise, although there are some similarities to various other versions.[65] The text discusses numeration, which it refers to as 'progression,' only briefly before proceeding to present addition, subtraction, duplation, mediation, multiplication, and division. Beyond merely presenting the forms of the Indo-Arabic digits, as in most texts, there are 10 verses explicating the shapes of the numerals. For example, four is described as "un 'o' qui .ii. piés" [an 'o' with two feet].[66]

A second French work from the 13th century is, for the most part, a prose translation of the *Carmen de algorismo*.[67] Although it differs from the latter in its list of operations, omitting extraction of roots and so listing only six arithmetic operations (referred to as *parties*),[68] the treatise in fact ends with a discussion of both square and cube roots. The work is very short, so much so that Louis Karpinski claims it to be "too brief to be useful in teaching the new arithmetic."[69]

By the 14th century, vernacular translations began to appear in other languages as well. For example, Jacob of Florence wrote the *Tractatus algorismi* in Tuscan in 1307.[70] The *incipit*, in Latin, indicates that the treatise will cover nine operations—namely, numeration, addition, subtraction, mediation, duplation, mutiplication, division, progression, and extraction of roots. In actuality, it considers only numeration and mutiplication and division of integers and fractions, along with a chapter on algebra.[71] Another example is a fragment of an English translation of a gloss

---

63 Edward Grant, ed., *A Source Book in Medieval Science* (Cambridge, MA: Harvard University Press, 1974), 101.

64 Karpinski, "A Thirteenth Century Algorism."

65 Karpinski, "A Thirteenth Century Algorism," 46.

66 Karpinski, "A Thirteenth Century Algorism," 58–59.

67 Victor Mortet, "Le plus ancien traité français d'algorisme," *Bibliotheca mathematica* 3, no. 9 (1908): 55–64.

68 Mortet, "Le plus ancien traité français d'algorisme," 61.

69 Karpinski, "A Thirteenth Century Algorism," 45.

70 Jens Høyrup, *Jacopo da Firenze's Tractatus Algorismi and Early Italian Abbacus Culture* (Basel: Birkhäuser, 2007).

71 Louis C. Karpinski, "The Italian Arithmetic and Algebra of Master of Jacob of Florence, 1307," *Archeion* 11, nos. 2–3 (1929): 170.

on the *Carmen de algorismo*.[72] The short fragment treats only numeration, but evidently is a translation of the same gloss as that found in *The Crafte of Nombrynge*, a 15th-century text in English.[73] In addition, Benedict lists vernacular texts in Hebrew and Greek.[74]

1.3 The *Algorismus* in Iceland

A version of the Indian calculus in Old Norse has been found in four separate manuscripts in Iceland. AM 544 4to (folios 90r to 93r),[75] GKS 1812 4to (folios 13v to 16v),[76] and AM 685 d 4to (folios 24v to 29r)[77] contain the full text, while AM 736 III 4to contains a fragment of the text in folios 4r to 4v.[78] The oldest of these is thought to be the copy in AM 544, a part of *Hauksbók* (a collection of texts compiled by Haukr Erlendsson, an Icelandic lawman), possibly written in the period 1306–1308.[79]

AM 685 d is much younger than either AM 544 or GKS 1812, most likely dating from the second half of the 15th century.[80] Even younger, AM 736 III is likely to be from the middle of the 16th century.[81]

72 James Orchard Halliwell, ed., "A Treatise on the Numeration of Algorism," in *Rara Mathematica* (London: John William Parker, 1839), 29–31; Robert Steele, ed., "A Treatise on the Numeration of Algorism," in *The Earliest Arithmetics in English* (London: Oxford University Press, 1922), 70–71.

73 Steele, *The Earliest Arithmetics in English*, v–vi; Robert Steele, ed., "The Crafte of Nombrynge," in *The Earliest Arithmetics in English* (London: Oxford University Press, 1922), 3–32.

74 Benedict, *A Comparative Study of the Early Treatises*, 17–18; Benedict also lists a version in Hebrew from the 12th century: Benedict, *A Comparative Study of the Early Treatises*, 10–11.

75 AM 544 4to (Copenhagen: Institut for Nordiske Studier og Sprogvidenskab), accessed December 10, 2025, https://sprogsamlinger.ku.dk/q.php?p=ds/hjem/mapper/9198.

76 GKS 1812 4to (Copenhagen: Institut for Nordiske Studier og Sprogvidenskab), accessed December 10, 2025, https://sprogsamlinger.ku.dk/q.php?p=ds/hjem/mapper/35612.

77 AM 685 d 4to (Copenhagen: Institut for Nordiske Studier og Sprogvidenskab), accessed December 10, 2025, https://sprogsamlinger.ku.dk/q.php?p=ds/hjem/mapper/11658.

78 AM 736 III 4to (Copenhagen: Institut for Nordiske Studier og Sprogvidenskab), accessed December 10, 2025, https://sprogsamlinger.ku.dk/q.php?p=ds/hjem/mapper/12327.

79 Otto B. Bekken and Marit Christoffersen, "Algorismus," in *Medieval Scandinavia: An Encyclopedia (Encyclopedias of the Middle Ages)*, ed. Phillip Pulsiano and Kirsten Wolf (Oxon: Routledge, 2016), 8; Gunnar Harðarson and Stefán Karlsson, "Hauksbók," in *Medieval Scandinavia: An Encyclopedia (Encyclopedias of the Middle Ages)*, ed. Phillip Pulsiano and Kirsten Wolf (Oxon: Routledge, 2016), 271.

80 "ONP: Dictionary of Old Norse Prose (Manuscripts)," Department of Nordic Studies and Linguistics University of Copenhagen, accessed December 10, 2025, https://onp.ku.dk/onp/onp.php?m.

81 "ONP: Dictionary of Old Norse Prose (Manuscripts)."

The *Algorismus* in Old Norse is clearly based on a paraphrastic translation of the *Carmen de algorismo*.[82] However, what sets it apart from the *Carmen de algorismo*, and other known versions of the Indian calculus of the time, is the final section devoted to a neoplatonic description, based on medieval glosses of the *Timaeus* of Plato, of the creation of the four elements. Although the immediate source for this section remains unknown, we explore the history of the development of this description of the classical elements in chapter 3 to help understand how these ideas found their way into a 14th-century treatise on arithmetic.

1.4 A Geometric Diagram in GKS 1812

Folio 13r of GKS 1812, immediately preceding the *Algorismus*, contains a short text concerning the geometry of circles, squares, and triangles inscribed in one another. Immediately below the text is a diagram illustrating the figures and computations described in the text, as shown in figure1.1.[83] The text appears along with the *Algorismus* in two other manuscripts: AM 685 d (folios 29r to 30r) and AM 736 III (folios 2r to 2v), as well as in the slightly older AM 415 4to (folio 9v, thought to be from around 1310).[84]

The computations described in this passage are commonplace, and so it is hard to pinpoint an exact source. However, the text immediately preceding this discussion in AM 415 describes the computations found in the *Commentarii in somnium Scipionis* of Macrobius, written around the turn of the 5th century, in regard to the circumference and radius of the earth.[85] Moreover, just preceding this section, Macrobius describes the relationship between the diameter and the circumference of a circle—that is, the circumference is equal to $\frac{22}{7}$ times the diameter—and provides the same example as found in the text we are considering (namely, that a circle of diameter 7 has a circumference of 22). The same computation of the diameter of the earth, given the circumference of the earth, is found in the *Tractatus de sphera* of

82 P. A. Munch, "Algorismus, eller Anviisning til at kjende og anvende de saakaldte arabiske Tal, efter Hr. Hauk Erlendssons Codex," in *Annaler for nordisk oldkyndighed og historie* (Copenhagen: J. D. Qvist, 1848), 374; *Hauksbók*, ed. Finnur Jónsson and Eiríkur Jónsson (Copenhagen: Thieles Bogtrykkeri, 1892–96), CXXXII.

83 ©The Arni Magnusson Institute for Icelandic Studies. Photograph: Sigurður Jónsson

84 "ONP: Dictionary of Old Norse Prose (Manuscripts)"; "AM 415 4to," in *Handrit.is* (Reykjavík: The Árni Magnússon Institute for Icelandic Studies), accessed March 2, 2026, https://handrit.is/manuscript/view/is/AM04-0415.

85 Macrobius, *Commentarii in Ciceronis somnium Scipionis*, ed. Ludwig von Jan (Quedlinburg and Leipzig: Godofredius Bassius, 1848), 556; Macrobius, *Commentary on the Dream of Scipio*, trans. William Harris Stahl (New York: Columbia University Press, 1990), 172.

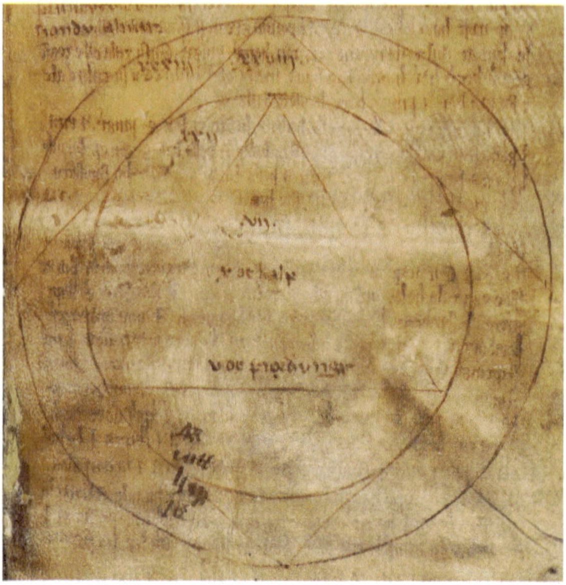

Fig. 1.1: Diagram of a circle with inscribed geometrical figures from GKS 1812, f. 13r

John of Sacrobosco.[86] As Natanael Beckman and Kristian Kålund indicate in their edition of the text,[87] some of the calculations may originate in the works of Gerbert of Aurillac from the latter part of the 10th century.[88]

Still, as Gustaf Eneström noted in 1898, it would be of interest to know if this work is original to whomever compiled the results, or if it is translation of a part of some previous geometric treatise:

> Il serait intéressant de connaître si le fragment a été traduit textuellement de quelque traité de géométrie, ou s'il est une compilation libre. Probablement les expressions »ferskeyta jafn» pour »carré», et »thriskeyta jafn» pour »triangle equilatéral» sont inventées par le traducteur

86 Iohannis de Sacro Bosco, "Tractatus de sphera," in *The Sphere of Sacrobosco and Its Commentators*, ed. Lynn Thorndike (Chicago: The University of Chicago Press, 1949), 85; John of Sacrobosco, "The Sphere of Sacrobosco," in *The Sphere of Sacrobosco and Its Commentators*, ed. and trans. Lynn Thorndike (Chicago: The University of Chicago Press, 1949), 123.

87 Natanael Beckman and Kristian Kålund, eds., *Alfræði Íslenzk*, vol. 2 (Copenhagen: S. L. Møllers Bogtrykkeri, 1916), 232.

88 Gerbert of Aurillac, "Geometria Gerberti," in *Oeuvres de Gerbert*, ed. Alexandre Olleris (Paris: Dumoulin, 1867), 403–70; Gerbert of Aurillac, "Epistola ad Adalboldum," in *Oeuvres de Gerbert*, ed. Alexandre Olleris (Paris: Dumoulin, 1867), 477–78.

ou compilateur; d'autre part les mots »horn» pour »angle» et »ummæling» pour »périmètre» ne sont guère des termes nouveaux.[89]

[It would be interesting to know if the fragment that has been translated directly from some treatise on geometry, or if is an original compilation. Probably the expressions 'ferskeyta jafn' for 'square,' and 'thriskeyta jafn' for 'equilateral triangle' are invented by the translator or the compiler; furthermore, the words 'horn' for 'angle' and 'ummæling' for 'perimeter' are hardly new terms.]

89 Gustaf Eneström, "Sur quelques propositions de planimétrie énoncées dans un manuscrit norvégien du 14e siècle," *Bibliotheca Mathematica* 12 (1898): 21–22.

2 The Algorithms of the *Algorismus*

The *Algorismus* in Old Norse states that there are seven branches (*greinar*) to the art known as *algorismus*. These are, as in the *Carmen de algorismo*, the operations of addition, subtraction, duplation, mediation, multiplication, division, and the extraction of roots. The extraction of roots has, in turn, two branches: finding square and cube roots. There are, in fact, two additional algorithms briefly described in the opening lines: a description of how to express numbers in the Indo-Arabic representational system—that is, numeration—and a method to identify even and odd numbers. The methods employed in the *Algorismus* are similar to those still taught in schools today, although the implementations are adapted for use with the dust boards and wax tablets of the day, which allowed for easily erasing and replacing intermediate calculations with updated ones, rather then the pen and paper calculations introduced in later texts on arithmetic.

2.1 Numeration

The *Algorismus* contains a basic description of the Indo-Arabic numeration system for writing numbers. We are told that, when written alone, each of the nine characters 1, 2, 3, 4, 5, 6, 7, 8, and 9 denotes its value—specifically, one through nine—and every shift of place to the left increases its value ten-fold.[1] Additionally, a tenth character, the cipher, corresponding to our modern 0, is interpreted as nothing more than a symbol, denoting nothing in itself but serving a purpose in holding a place for the signification of the other digits in the number.

The text describes three types of numbers. A number which is less than ten is called a *fingr*–that is, 'finger'—a direct translation of the Latin *digitus*. A number which begins with a cipher—namely, a multiple of ten—is called a *líðr*, which literally means 'joint,' a direct translation of the Latin *articulus*. The words *digitus* and *articulus* are indicative of an ancient form of calculating with fingers.[2] Finally, a number which is the sum of a *líðr* and a *fingr* is called a *samsett tala*, which is literally 'composite number,' a translation of *compositus numerus*. As 'composite number' has a more common mathematical meaning of denoting a positive integer, greater than 1, which is the product of two smaller integers, *samsett tala* is

1 The author notes that the system is like the system for writing in Hebrew—namely, right-to-left rather than left-to-right.

2 See, for example, Menso Folkerts, "Early Texts on Hindu-Arabic Calculation," *Science in Context* 14, nos. 1–2 (2002): 13.

 | https://doi.org/10.1515/9781501522376-002

translated here as 'compound number.' Three examples are provided: 8 for a *fingr*, 70 for a *líðr*, and 65 for a *samsett tala*. This is significantly simpler than the example of 1,180,703,051,492,863 provided in the *Dixit Algorizmi*.[3]

2.2 Even and Odd Numbers

The standard definition of an even number is a number which is divisible by two. For example, Euclid says in Book VII, Definition 6, that "[a]n even number is that which is divisble into two equal parts."[4] In fact, the Latin word for 'even' is *par*, which has the basic meaning of 'equal.'[5] The *Algorismus* uses *jafn*, a direct translation for *par*. The use of the Indo-Arabic numbers provides a simple way to distinguish between even and odd numbers—namely, even numbers begin with one of the digits 0, 2, 4, 6, or 8—and odd numbers with one of the digits 1, 3, 5, 7, or 9. As the author of the *Algorismus* considers the cipher to be just a place holder, not denoting anything itself, and one not to be a number, the text has to consider several conditions. First, an even number is one which is either a multiple of ten (hence any number which begins with a cipher) or begins with an even digit (those being 2, 4, 6, and 8). An odd number is said to be one which begins with an odd digit, those being 3, 5, 7, and 9. It is stated that one is neither an even nor an odd number (since it is not considered a number at all), but there is no statement about other numbers which begin with the digit 1.

2.3 Arithmetic Operations

2.3.1 Addition and Subtraction

For addition, the algorithm begins with specifying that the two numbers be written down, one above the other, aligned on the right, with the smaller of the two numbers written on the lower line. The process is then identical to the one still taught in schools today, except that the result gradually replaces the upper number rather than, as in modern instruction, being written below the two numbers. That is, corresponding digits are added, one-by-one, from right to left. If the result is a digit, the new digit simply replaces the old digit in the upper number. If the result is an

3 Allard, *Le Calcul Indien (Algorismus)*, 6.

4 Euclid, *The Thirteen Books of Euclid's Elements*, trans. T. L. Heath, vol. 2 (Cambridge: Cambridge University Press, 1908), 277.

5 Similarly for the Greek word, 'ἄρτιος,' used by Euclid.

article or a compound number, the article is added to the next digit to the left and the digit (or cipher) is written in place of the the original digit.

The text does not mention what changes, if any, should be made to the lower number. An example given by Peter of Dacia shows the digits of the lower number replaced, one-by-one as the process proceeds, by ciphers,[6] while the commentary in *The Crafte of Nombrynge* instructs the reader to let the lower number stay as it is.[7] It would also be reasonable to suppose that the lower number was gradually erased as the digits were used, which is the notation we will use in the example below.

For example, to add 2078 to 1325, the reader is instructed to first write the two numbers one under the other, with the smaller number below and aligned to the right:

$$\begin{array}{cccc} 2 & 0 & 7 & 8 \\ 1 & 3 & 2 & 5 \end{array}.$$

Starting from the right, we add the lower digit to the upper digit. If the result is a digit, we simply replace the upper digit (in this example, 8) with the new digit. But here the sum of 8 and 5 is 13, so we are told to replace the 8 with the 3 from the 13 and add the 1 from the 13 to the 7 in the next place. Hence we now have

$$\begin{array}{cccc} 2 & 0 & 8 & 3 \\ 1 & 3 & 2 & \end{array}.$$

We now move to the next digit of the lower number, and repeat the procedure. This time we add 2 to 8, which gives us 10, which is an article. The text tells us we should replace the 8 in the upper number with a cipher and add the 1 of the article to the next place. Since the next place is a cipher, we simply replace it with the 1. So we now have

$$\begin{array}{cccc} 2 & 1 & 0 & 3 \\ 1 & 3 & & \end{array}.$$

Proceeding to the next digit, we add 3 to 1 to obtain 4, which, since it is a digit, simply replacess the 1. This gives us

$$\begin{array}{cccc} 2 & 4 & 0 & 3 \\ 1 & & & \end{array}.$$

Finally, reaching the end, we add 1 to 2, which is again a digit, and so we obtain the final result

$$3 \quad 4 \quad 0 \quad 3\ .$$

For subtraction, the algorithm begins with writing the numbers down as with addition, with the smaller number written below the larger. Again, the process is much

6 Petrus de Dacia, "Expositio super Algorismum," 100–101.
7 Steele, "The Crafte of Nombrynge," 9–10.

the same as with the modern one: the digits of the lower number are subtracted, one-by-one, from the digits of the upper number, proceeding from right to left. As with addition, the upper number is gradually replaced by the digits of the result. The text again does not mention any changes to the lower number. The examples given by Peter of Dacia and in *The Crafte of Nombrynge* show the lower number unchanged throughout the process, but it is again reasonable to suppose that the lower number is gradually erased as the digits are used.

For example, to subtract 1872 from 2015, we begin by writing the numbers down as in the previous example, although it is now essential that the smaller number be written below the larger. So we start with

$$\begin{array}{cccc} 2 & 0 & 1 & 5 \\ 1 & 8 & 7 & 2 \end{array}.$$

Starting with the first digits on the right, we are instructed to take the lower digit from the upper digit, replacing the upper digit with the result if something remains (that is, if the result is nonzero) and with a cipher if nothing remains. For this example, 2 from 5 leaves 3, so we now have

$$\begin{array}{cccc} 2 & 0 & 1 & 3 \\ 1 & 8 & 7 & \end{array}.$$

But a problem arises if a digit in the top number is smaller than the corresponding digit in the lower number. In that case, we are told to take one away from the next digit of the top number and add it, as 10, to the digit in the top number that we are working with. Again, there is a potential problem if the next digit is actually a cipher. In that case, we must find the first digit which is not a cipher to the left of the current digit, reduce it by one, and replace all the ciphers with nines. In this example, that means, for the second step, we add 10 to the digit above the 7, making it 11, and replace the 2 and 0 to its left with 1 and 9. We then subtract the 7 of the lower number from 11, to obtain 4, and so the completed step looks like

$$\begin{array}{cccc} 1 & 9 & 4 & 3 \\ 1 & 8 & & \end{array}.$$

Proceeding to the next step, taking 8 from 9 leaves 1, so we have

$$\begin{array}{cccc} 1 & 1 & 4 & 3 \\ 1 & & & \end{array}.$$

And so, finally, subtracting 1 from 1 will leave nothing. Although the text says to write a cipher whenever this happens, in practice ciphers to the left of a number would be omitted. So the final result is now

$$\begin{array}{ccc} 1 & 4 & 3 \end{array}.$$

2.3.2 Duplation and Mediation

The algorithms for duplation and mediation are similar to each other, although the former proceeds from the left and the latter from the right. For duplation, one successively doubles the digits of the given number from left to right; for mediation, one successively halves the digits of the number from right to left.

To double a number, we are told to first write the number. For example, we might start with

$$4 \quad 7 \quad 9 \quad 3\ .$$

Starting from the left, we double the first digit, replacing the 4 with an 8 to obtain

$$8 \quad 7 \quad 9 \quad 3\ .$$

Proceeding to the next digit, doubling 7 gives us 14. Since this is a compound number, we are told to write the digit 4 in the same place, but to add the 1 to the preceding 8, as we would in addition. Hence we now have

$$9 \quad 4 \quad 9 \quad 3\ .$$

Moving to the next digit, doubling 9 gives us 18, so we replace the 9 with an 8 and add 1 to the preceding 4 to obtain 5. So we now have

$$9 \quad 5 \quad 8 \quad 3\ .$$

Finally, we double 3, giving us 6 in the rightmost place, and our final answer is

$$9 \quad 5 \quad 8 \quad 6\ .$$

The section has an extra instruction for the case where a *semis*—that is, the symbol ɣ', denoting $\frac{1}{2}$—is written over the first digit on the right. This symbol would appear there only if the number being doubled was the result of some odd number having previously been halved.[8] That procedure is described in the next section on mediation, so the insertion here may be an indication that the order of presentation of the operations of duplation and mediation were reversed in some earlier text from which the *Algorismus* and the *Carmen de algorismo* were derived. Here we are told that, in the process of duplation, if a *semis* is written over the first digit of the original number, then one should be added to the first digit after it has been doubled.

8 The text says *even* number, but this is clearly a scribal error, which occurs as well in AM 544 but is stated correctly in AM 685 d.

Mediation begins from the right. If the first digit is even, it is divided in half. But if it is odd, the instructions are to subtract one from that digit, divide the resulting even number in half, and denote the remainder of $\frac{1}{2}$ by writing a ♈, a *semis*, over the last digit of the result. There appears to be some missing text in regard to the case if the first digit is a one, but there is an indication that, for that case, a cipher should be written where the one was.

Different versions of the Indian calculus denote this remainder of $\frac{1}{2}$ in various ways. The *Carmen de algorismo* instructs the reader to use a one rather than a *semis*, although *The Crafte of Nombrynge* uses a "w" in its translation.[9] Other texts, such as the *Dixit Algorizmi*, the *Liber ysagogarum Alchorismi*, the *Liber Alchorismi*, and the *Liber pulueris*, use a sexagesimal system—that is, a system using minutes and seconds—to denote fractions, and so denote the remainder as 30 out of 60. For example, the two 12th-century texts edited by Karpinski describe writing 30 under the last digit (representing the fraction $\frac{30}{60}$) to denote the remaining $\frac{1}{2}$.[10] The *Algorismus vulgaris*, as well as the commentary of Peter of Dacia, mentions various ways for denoting the remainder, including writing the sexagesimal minutes, writing a 1, or writing a figure for $\frac{1}{2}$.[11] Moreover, the latter four texts all discuss mediation before duplation.

Mediation then proceeds through the rest of the digits in a similar way, with one modification: For odd digits, the procedure is to halve the even digit one less than the given digit and then add five to the previous place—namely, one position to the right. For example, beginning with

7 4 5

the first step would be to subtract 1 from 5, divide the resulting 4 in half, and then replace the 5 with a 2 and a raised *semis*, giving us

7 4 2♈ .

Next, divide the middle 4 in half, obtaining

7 2 2♈ .

For the final step, subtract 1 from 7, divide the resulting 6 in half, and add the remainder 5 to the previous 2, resulting in the final outcome

3 7 2♈ .

9 Steele, "Carmen de algorismo," 74; Alexander de Villedieu, "Carmen de algorismo," 76; Steele, "The Crafte of Nombrynge," 17.

10 Karpinski, "Two Twelfth Century Algorisms," 400.

11 Iohannis de Sacro Bosco, "Algorismus," 182; Petrus de Dacia, "Expositio super Algorismum," 109.

2.3.3 Multiplication and Division

We will illustrate the algorithm for multiplying two integers using an example. The text itself contains none, but the technique is easier to grasp with an example than with a general description. Although similar to the modern method, the operations begin from the left instead of from the right and the intermediate results gradually replace the number on the top line rather than being written below. To begin the algorithm, the two numbers to be multiplied are written, one under the other, but, unlike with addition and subtraction, the lower number is written with its right-most digit beneath the leftmost digit of the upper number. So, for example, to find 875×432, we first write

$$\begin{array}{ccccc} & & 8 & 7 & 5 \\ 4 & 3 & 2 & & \end{array}.$$

The next step is to multiply the lower number, 432, by the left-most digit of the upper number, 8, one digit at a time, starting with the left-most digit, 4, while writing the result directly above the lower number. Thus, since 8×4 is 32, we first write

$$\begin{array}{cccccc} 3 & 2 & & 8 & 7 & 5 \\ & 4 & 3 & 2 & & \end{array}.$$

Next, 8×3 is 24, so we write the digit 4 above lower 3, and add the article 2 to the 2 above the 4. Now we have

$$\begin{array}{cccccc} 3 & 4 & 4 & 8 & 7 & 5 \\ & 4 & 3 & 2 & & \end{array}.$$

Then 8×2 is 16, so replace the 8 in the upper number by a 6 and add 1 to the 4 above the 3. At this point, after multiplying 432 by 8, we have

$$\begin{array}{cccccc} 3 & 4 & 5 & 6 & 7 & 5 \\ & 4 & 3 & 2 & & \end{array}.$$

The next step requires shifting the lower number one place to the right, and repeating the previous procedure to multiply 432 by 7, digit by digit. So, starting with

$$\begin{array}{cccccc} 3 & 4 & 5 & 6 & 7 & 5 \\ & & 4 & 3 & 2 & \end{array},$$

we multiply 7×4, which is 28, to find

$$\begin{array}{cccccc} 3 & 7 & 3 & 6 & 7 & 5 \\ & & 4 & 3 & 2 & \end{array}.$$

Then 7×3 is 21, giving us

$$\begin{array}{cccccc} 3 & 7 & 5 & 7 & 7 & 5 \\ & & 4 & 3 & 2 & \end{array}.$$

Next 7×2 is 14, and so the digit 4 replaces the rightmost 7 in the upper number and 1 is added to the 7 to its left. Thus, after two multiplications, we have

$$\begin{array}{cccccc} 3 & 7 & 5 & 8 & 4 & 5 \\ & & 4 & 3 & 2 & \end{array} .$$

We now shift the lower number one place to the right, giving us

$$\begin{array}{cccccc} 3 & 7 & 5 & 8 & 4 & 5 \\ & & & 4 & 3 & 2 \end{array} ,$$

and multiply 5 times 432. Now 5×4 is 20, so we have

$$\begin{array}{cccccc} 3 & 7 & 7 & 8 & 4 & 5 \\ & & & 4 & 3 & 2 \end{array} .$$

And 5×3 is 15, giving us

$$\begin{array}{cccccc} 3 & 7 & 7 & 9 & 9 & 5 \\ & & & 4 & 3 & 2 \end{array} .$$

Finally, 5×2 is 10, so the 5 in the upper number is replaced by a cipher, 0, and 1 is added to the place to its left, resulting in two more ciphers, giving us the final result

$$\begin{array}{cccccc} 3 & 7 & 8 & 0 & 0 & 0 \\ & & & 4 & 3 & 2 \end{array} .$$

Thus we have 875×432 is 378,000.

To efficiently use the multiplication algorithm, one needs to be able to easily find products of the digits 1 through 9. In modern schools, this is taught through memorization of the table of the products. This is also seen in some versions of the Indian calculus. For example, the *Liber ysagogarum Alchorismi* contains such a table.[12] Other texts describe various rules for aiding in the computation of these products. The treatise from the Salem Abbey presumes knowledge of the products of the digits 1 through 5 and then provides rules for finding the other products.[13] For example, if both a and b are digits greater than 5, then

$$ab = 10(b - (10 - a)) + (10 - a)(10 - b)$$

allows one to find ab knowing only the products of the digits 1 through 5. For example,

$$8 \times 7 = 10(7 - (10 - 8)) + (10 - 8)(10 - 7) = (10 \times 5) + (2 \times 3) = 56.$$

12 Allard, *Le Calcul Indien (Algorismus)*, 27.

13 Cantor, “Über einen Codex des Klosters Salem,” 5; Benedict, *A Comparative Study of the Early Treatises*, 89.

Also, multiples of 9 are found easily using, for any digit a,

$$9a = 10a - a. \tag{2.1}$$

For example,

$$9 \times 4 = 10 \times 4 - 4 = 36.$$

The *Carmen de algorismo* and the *Algorismus* provide only one aid for finding products of digits. This is described as follows in the *Algorismus*:

> þar nœſt ſkalt þu hugſa huerſſu mikít ena meiri figuru ſkortír a tíu. þa er þu uilt marghfalda. Ok ſua marghar ein<í>ngar ſem aſkortír a tíu. ſua opt ſkalt þu ena mínní toluna þaa er þu vilt margfaldda taka af tíghum hennar. ok at þu ſkilir þetta margfaldda ſeau ok níu. Nío ſkortír eín a tío. þuí tak þu eína ſeau af ſeautighum. þa uerda eftir þrír ok ſextíghí. þat eru ſeau ſínnum níu. At ſlíku ſkapi mat þu aðrar tolur rǿyna.[14]
>
> [Next you shall consider how much the larger figure which you want to multiply lacks from ten. And as many units as are lacking from ten, that is how often you should take the smaller number, that you want to multiply, from that number of tens. So that you understand this, multiply seven and nine. Nine is one less than ten, therefore you take one seven away from seventy, and then sixty-three remains. That is seven times nine. In the same manner you can attempt other numbers.]

That is, to multiply two digits, consider the difference between the larger digit and ten, then subtract the smaller number that many times from the smaller digit number of tens. Although the *Carmen de algorismo* does not provide an example, the *Algorismus* gives one of the simplest type:

$$7 \times 9 = (10 \times 7) - 7(10 - 9) = 70 - 7 = 63,$$

The method described here is a general version of (2.1): For any integers a and b between 1 and 9,

$$ab = 10a - a(10 - b), \tag{2.2}$$

although the example given would also follow from (2.1). Although the text specifies that the formula should be used with a less than b, no further indication is given as to when (2.2) is more efficient than a direct calculation. For example, (2.2) is perhaps useful in finding 7×8, but not for something like 3×5.

For division, the numbers are arranged differently than our current method for long division, but the calculations are identical. The divisor is written below the number being divided (aligned to the left), the successive parts of the quotient are written above the top number, and the number being divided is successively

14 GKS 1812 4to, f. 14v, ll. 4–10.

dimished by the intermediate products as the algorithm proceeds. For example, to divide 87,453 by 272, we first write the two numbers as such

$$\begin{array}{ccccc} 8 & 7 & 4 & 5 & 3 \\ 2 & 7 & 2 & & \end{array}.$$

Then we note that 272 divides into 874 three times. Multiplying 272 by 3 gives us 816, which is then subtracted from 874, leaving 58. Writing the digit 3 above the upper number, aligned with the right-most digit of the lower number, and replacing 874 with 58 gives us

$$\begin{array}{ccccc} & & 3 & & \\ & 5 & 8 & 5 & 3 \\ 2 & 7 & 2 & & \end{array}.$$

The instructions now say to shift the divisor 272 one space to the right and repeat the process with

$$\begin{array}{cccc} & 3 & & \\ 5 & 8 & 5 & 3 \\ 2 & 7 & 2 & \end{array}.$$

This time we see that 272 divides 585 two times. Writing the 2 above the upper number, aligned with the right-most digit of the divisor, and subtracting $272 \times 2 = 544$ from 585, which leaves 41, we now have

$$\begin{array}{cccc} & 3 & 2 & \\ & 4 & 1 & 3 \\ 2 & 7 & 2 & \end{array}.$$

We again shift 272 one space to the right, giving us

$$\begin{array}{ccc} 3 & 2 & \\ 4 & 1 & 3 \\ 2 & 7 & 2 \end{array}.$$

Then we see that 272 divides 413 only once. Since subtracting 272 from 413 leaves 141, we add the 1 to the end of our quotient and replace 413 with the remainder 141. This is the final step, and we have

$$\begin{array}{ccc} 3 & 2 & 1 \\ 1 & 4 & 1 \\ 2 & 7 & 2 \end{array}.$$

We read this as saying that 87,453 divided by 272 is 321, with a remainder of 141. That is,

$$272 \times 321 + 141 = 87{,}453,$$

a computation the text recommends to check that the division has been performed correctly.

2.4 Finding Roots

2.4.1 Square Roots

The algorithm for finding square roots in the *Algorismus* was known in China at least as far back as the 3rd century CE and continued to be taught in schools well into the 20th century.[15] Although not the most efficient method for finding square roots, it has the virtue of producing the exact digits of the root, one by one, starting from the left, much in the way the division algorithm produces successive digits of a quotient.

The method relies on two basic observations. The first concerns the number of digits in the number N for which we wish to find the square root. Note that the square of a single digit is a number with 1 or 2 digits, the square of a number with two digits is a number with 3 or 4 digits, the square of a number with three digits is a number with 5 or 6 digits, and so on. It follows that the number of digits of the square root r of N will depend on pairing the digits of N, starting from the right. If N has an even number of digits, this process will end with the two rightmost digits of N. If N has an odd number of digits, it will end with the single rightmost digit of N. This is the case treated explicitly in the text, beginning with instructions to find the largest digit whose square is less than or equal to the first digit of the given number N. For the case where N has an even number of digits, the section ends with the comment:

> En ef fyſti ſtadir þeirrar tolu er þu ritadir uar iafn þa fín fíngir undir nœſtu figuru ok margfalða a ſomu leid.[16]
>
> [If the first place of that number which you wrote was even, then find a digit under the next figure and multiply in the same way.]

In this case the first step is to find the largest digit whose square is less than or equal to the first two digits of N.

The second observation is that if x is an approximation for the square root r of N, with $x \leq r$, and if we let $y = r - x$, then

$$N = (x + y)^2 = x^2 + 2xy + y^2.$$

And so

$$N - x^2 = 2xy + y^2.$$

15 Jean-Luc Chabert et al., *Histoire d'algorithmes: Du caillou à la puce*, ed. Jean-Luc Chabert (Paris: Belin, 1993), 234–35.

16 GKS 1812 4to, f. 15v, ll. 14–16.

Hence if x is an initial under approximation of the root r, then we may improve the approximation by finding a value of y for which

$$2xy + y^2 \le N - x^2. \qquad (2.3)$$

The algorithm specifies how to use this inequality to successively create the digits of r. That is, if r has m digits, we take x to be the first digit of r followed by $m - 1$ zeros and then look for the largest value of y, in the form of a digit followed by $m-2$ zeros, for which the inequality (2.3) holds. In practice, the computations are simplified by working with only a part of the given number N at a time, keeping track of the intermediate numbers in their appropriate places. The procedure continues until the digits of N have been exhausted. If some part of N remains, then N is not an exact square of an integer and what remains is treated as a remainder, as in division.

The main work of the text in describing the algorithm is to use the Indo-Arabic notational system to keep track of the positions of the intermediate calculations. Not having any form of symbolic representation for these calculations, the terms 'duple' and 'subduple' are introduced, the former denoting the double of a number and the latter the number itself.

For example, to find the square root r of 39,340, we first note that the number has an odd number of digits. Grouping the digits in pairs, starting from the right, there are three groups, with the leftmost group consisting of just the digit 3. So r will have three digits, which we will denote a, b, and c. That is, $r = (a \times 10^2) + (b \times 10) + c$. Now the first step is to find the largest digit a for which $a^2 \le 3$. It follows that $a = 1$. Subtracting $a^2 = 1$ from 3 leaves 29,340. Next, considering the leading digit 2 with the next two digits, we look for the largest digit b for which

$$2(a \times 10)b + b^2 \le 293.$$ [17]

Since $(2 \times (1 \times 10) \times 9) + 81 = 261$, we have $b = 9$. Subtracting 261 from 293 leaves 32. For the final step, we consider the leading 32 with the final two digits and seek the largest digit c for which

$$2(((a \times 10) + b) \times 10)c + c^2 \le 3{,}240.$$

With $a = 1$ and $b = 9$, $(a \times 10) + b = 19$, this becomes

$$(2 \times 190)c + c^2 \le 3{,}240.$$

Now if $c = 8$, we have

$$(380 \times 8) + 64 = 3{,}104,$$

17 The digit a is multiplied by 10 since it stands one place to the left of b.

but with $c = 9$, we obtain

$$(380 \times 9) + 81 = 3{,}501.$$

Hence $c = 8$, and our desired square root is $r = 198$, which leaves a remainder of $3{,}240 - 3{,}104 = 136$. Checking, we note

$$198^2 + 136 = 39{,}204 + 136 = 39{,}340.$$

In the notation of the *Algorismus*, the steps would appear as the following:

$$\begin{array}{ccccc} & & & & \\ 3 & 9 & 3 & 4 & 0 \end{array} \longrightarrow \begin{array}{ccccc} 1 & & & & \\ 2 & 9 & 3 & 4 & 0 \end{array} \longrightarrow \begin{array}{ccccc} 1 & & 9 & & \\ & 3 & 2 & 4 & 0 \end{array} \longrightarrow \begin{array}{ccccc} 1 & & 9 & & 8 \\ & & 1 & 3 & 6 \end{array}.$$

The text does not indicate where the numerous intermediate calculations are to be performed. Without the aid of symbolic representations for the intermediate steps, and without examples, it is hard to see how one could master the technique from reading the text.

2.4.2 Cube Roots

The algorithm for finding cube roots follows the same pattern as that for square roots, *mutatis mutandis*. We now group the digits of the given number N, starting from the right, in triplets. The leftmost group will have one to three digits, and we first look for a digit whose cube is less than or equal to that number. Having found an under approximation for the cube root r of N–namely, this digit followed by $m - 1$ zeros, where m is the number of groups of three digits in N—we call this approxmation x and let $y = r - x$. Now,

$$N = (x + y)^3 = x^3 + 3x^2y + 3xy^2 + y^3,$$

and so

$$N - x^3 = 3x^2y + 3xy^2 + y^3.$$

We then seek the largest number y, consisting of a digit followed by $m - 2$ zeros, for which

$$3x^2y + 3xy^2 + y^3 \le N - x^3.$$

As we saw in the previous example, the computations are simplified by working with only a part of the given number N at a time, keeping track of the intermediate numbers in their appropriate places. The procedure continues until the digits of N have been exhausted, perhaps terminating with some remainder. The text introduces the terms 'subtriple' and 'triple' to help keep track of a number and its multiple by three, respectively.

As with the algorithm for finding square roots, the technique is best understood through an example. So now consider finding the cube root s of 39,340. Grouping the digits in triplets, starting from the right, we have two groups, the leftmost group being 39. So s will have two digits, which we will denote d and e. That is,

$$s = (d \times 10) + e.$$

Now d will be the largest digit for which $d^3 \leq 39$. Hence $d = 3$. Subtracting $d^3 = 27$ from 39 leaves 12,340. For e, we need to find the largest digit for which

$$3(d \times 10)^2 e + 3(d \times 10)e^2 + e^3 \leq 12{,}340.$$

Now when $e = 4$, we have

$$(3 \times 30^2 \times 4) + (3 \times 30 \times 4^2) + 4^3 = 12{,}304$$

and when $e = 5$ we have

$$(3 \times 30^2 \times 5) + (3 \times 30 \times 5^2) + 5^3 = 15{,}875.$$

Hence $e = 4$. Thus the desired cube root is $s = 34$, with a remainder of $12{,}340 - 12{,}304 = 36$. That is,

$$39{,}340 = 34^3 + 36.$$

In the notation of the *Algorismus*, these steps would appear as the following:

$$\begin{array}{ccccccccccccccccc} & & & & & & & & 3 & & & & & 3 & & & 4 \\ 3 & 9 & 3 & 4 & 0 & \longrightarrow & 1 & & 2 & & 3 & 4 & 0 \longrightarrow & & & 3 & 6 \end{array}.$$

Once again, substantial intermediate calculations would need to be performed separately.

3 Neoplatonism and the Elements

> Hear first of all the four roots of all things:
> Zeus the gleaming, Hera who gives life, Aidoneus,
> And Nêstis, who moistens with her tears the mortal fountain.
>
> — Empedocles[1]

> But in fact the world was to be a solid, and solids must always be conjoined not by one middle term, but by two. So God inserted water and air between fire and earth, and made them all, as far as was possible, proportional to one another, air being to water as fire to air, and water to earth as air to water.
>
> — Plato[2]

> There is thus a possibility that the ancient dream of philosophers to connect all Nature with the properties of whole numbers will some day be realized.
>
> — Paul Dirac[3]

> For the reasons (*logoi*) that govern Nature are not receptive of the accuracy or the fixity of mathematicals.
>
> — Proclus[4]

The Old Norse *Algorismus* concludes with a numerical explication of the creation of the four classical elements which were, in the ancient and medieval world view, considered to be the constituents of the physical world known to our senses. This discussion of how the elements of earth, water, air, and fire came to be, and why there are just four, has not been found in any other manuscripts of the Indian calculus. However, there is a continuous thread of thought, dating back to at least Plato and running through a diverse set of ancient and medieval works, which connects the properties of the four elements with a geometrical progression from one cubic number to another. We can only speculate as to whether this passage was inserted by the original translator of the Old Norse *Algorismus* or a later scribe, and, if so, from what text it might have been borrowed, or if the section was already in the manuscript of some lost Latin *Algorismus* from which the translator worked. Nev-

1 André Laks and Glenn W. Most, eds., *Early Greek Philosophy*, vol. V (Cambridge, MA: Harvard University Press, 2016), 401.

2 Plato, *Plato: Timaeus and Critias*, trans. A. E. Taylor (London: Methuen & Co. Ltd., 1929), 32b, 29.

3 Paul Adrien Maurice Dirac, "The Relation Between Mathematics and Physics," *Proceedings of the Royal Society (Edinburgh)* 59 (1938–39): 129.

4 Proclus, *Commentary on Plato's Timaeus*, ed. and trans. Dirk Baltzly, vol. III (Cambridge: Cambridge University Press, 2007), 69.

 | https://doi.org/10.1515/9781501522376-003

ertheless, we can follow the thread to see where the sources might lie, and we can see why the text is not as out of place as it might at first appear to a modern reader.

3.1 The *Timaeus*

For some two and a half millennia, natural philosophers have held that all of the physical matter encountered by our senses in everyday life is composed from a small number of basic building blocks, the chemical elements. Although natural scientists now identify more than 90 such elements that occur naturally, for much of this time there were thought to be only four. In his *Metaphysics*, Aristotle credits the 5th century BCE philosopher Empedocles as being "the first to speak of four material elements."[5] Empedocles identified the "four roots of all things" with the four gods Zeus, Hera, Aidoneus, and Nêstis, who, in some order, signify the later traditional names of fire, earth, air, and water.[6]

However, it appears that Plato, writing in the age between Empedocles and Aristotle, was the first to attempt a mathematical justification for the existence, and necessity, of exactly four elements constituting the building blocks from which all of our natural world has been constructed. His argument hinges on the nature of our senses, with which we experience the world, and the three-dimensionality of the world which we encounter with those senses. Describing the creation of the physical world by the demiurge in the *Timaeus*, he begins with the assertion that anything bodily must be both visible and tangible. In the words of Timaeus,

> Being bodily, that which has come to be must be visible and tangible. Without fire nothing visible can come to be, nothing tangible without solidity, nothing solid without earth.[7]

That is, in order that we may interact with a world of three-dimensional bodies, there must be fire to make them visible and earth to make then solid and tangible. Further, these two elements, fire and earth, being three-dimensional and contrary to one another in every dimension (in a manner left for later philosophers to explicate), require intermediaries to bind them together. For Plato, this binding is perfected only with a geometric progression. As Timaeus argues,

---

5 Aristotle, "Metaphysics," in *The Complete Works of Aristotle: The Revised Oxford Translation*, ed. Jonathan Barnes, trans. W. D. Rouse, vol. 2 (Princeton, New Jersey: Princeton University Press, 1984), 985a32, 1558.

6 K. Scarlett Kingsley and Richard Parry, "Empedocles," in *The Stanford Encyclopedia of Philosophy*, Winter 2024, ed. Edward N. Zalta (Metaphysics Research Lab, Stanford University, 2024), accessed December 10, 2025, https://plato.stanford.edu/archives/win2024/entries/empedocles/.

7 Plato, *Plato: Timaeus and Critias*, 31b, 28.

> The fairest of all bonds is that which makes itself and the terms it binds together most utterly one, and this is most perfectly effected by a progression. For when of three integers, or volumes, or characters, the midmost is to the last as the first to itself, and, inversely, as the last to the midmost so the midmost to the first, then all will necessarily take one another's places, the midmost becoming first and last, and both the first and last midmost, and so, by substitution for one another, they form a perfect unity.[8]

From here, Timaeus reaches the conclusion that

> God inserted water and air between fire and earth, and made them all, as far as was possible, proportional to one another, air being to water as fire to air, and water to earth as air to water.[9]

To make sense of what Timaeus is claiming in reaching this conclusion, he must move the problem from one of physical bodies to one of mathematics. This is a move common in pythagorean thought and Plato seems comfortable in making the shift without comment. However, the question of the relationship between mathematics and the physical world of our senses is a deep one which has engendered much philosophical and scientific debate over the centuries. We will return to this question in a later chapter, but for now it is worth noting the warning of Proclus, a 5th-century neoplatonic philosopher whose commentary on the *Timaeus* we will consider below:

> But on the other hand, it is also necessary not to dwell [too much] on the mathematical [parts of the dialogue] as some do. For this too engenders false opinions in the audience in as much as they come to think of physical figures being the same thing as mathematical numbers.[10]

In other words, we need not identify objects of the physical world with those of the mathematical world, as a pythagorean might, in order to employ mathematics as a means to better understand how the objects of the physical world relate to one another.

The argument that Plato presents to reach his conclusion has long been a subject for debate among commentators on the *Timaeus*. In somewhat modern language, we might reconstruct this argument more or less along the following lines: First, we must recall that pythagoreans, and Greek mathematicians more generally, identified the product of two numbers, say a and b, with the rectangle, or at least with the area of the rectangle, with length a and width b. The product of three numbers, a, b, and c, was identified with a rectangular parallelepiped with length a, width b, and height c (and, hence, volume abc). For simplicity, we will call such

8 Plato, *Plato: Timaeus and Critias*, 31c–32a, 28.
9 Plato, *Plato: Timaeus and Critias*, 32b, 29.
10 Proclus, *Commentary on Plato's Timaeus*, 3.23, 69.

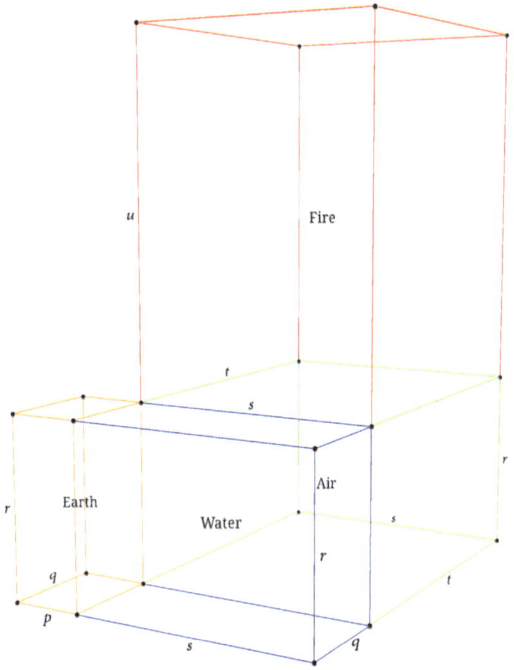

Fig. 3.1: Geometric progression between two boxes

a three-dimensional solid a box. Next, we identify the element earth with the solid having dimensions p, q, and r and the element fire with the solid having dimensions s, t, u. Moreover, we assume p, q, r, s, t, and u are all prime integers, with the latter three distinct from the first three, in keeping with the notion that prime numbers are the building blocks of all integers in a way somewhat analogous to the elements being the building blocks of all physical matter. Now we want to construct a continuous progression from the box with volume pqr to the box with volume stu. As shown in figure 3.1, we can do this by constructing two intermediary boxes, the first (that is, water) with volume sqr and the second (that is, air) with volume str.[11] Finally, we need these volumes to form a geometric progression, so we must have

$$\frac{pqr}{sqr} = \frac{sqr}{str} = \frac{str}{stu}.$$

11 This diagram is similar to the one found in Proposition 33 of Book XI of Euclid. For a 9th-century rendering, see: Euclid, "In Euclidis elementorum libri," in Manuscript Gr. 190.pt.2 (Vatican: Vatican Library, Digivatlib), f. 209r, accessed February 11, 2026, https://digi.vatlib.it/mss/detail/Vat.gr.190.pt.2; For a modern rendering, see: Euclid, *The Thirteen Books of Euclid's Elements*, trans. T. L. Heath, vol. 3 (Cambridge: Cambridge University Press, 1908), 342.

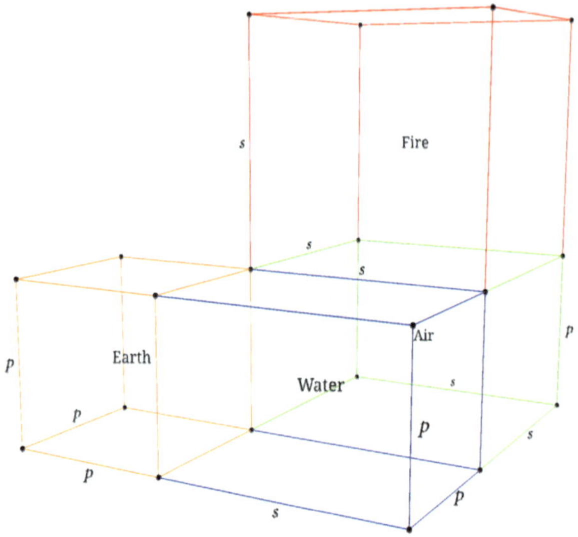

Fig. 3.2: Geometric progression between two cubes

Hence, we want

$$\frac{p}{s} = \frac{q}{t} = \frac{r}{u},$$

which implies that $pt = qs$ and $pu = rt$. Since the factors are all prime, and the first triple is distinct from the second, this in turn implies

$$p = q = r \text{ and } s = t = u.$$

Thus, earth is identified with p^3, water with p^2s, air with ps^2, and fire with s^3, as shown in figure 3.2. The resulting sequence, p^3, p^2s, ps^2, and s^3, forms a geometric progression, with the ratio of successive terms being $\frac{s}{p}$. The authors of later commentaries will take the primes p and s to be 2 and 3, the first two primes, resulting in the geometric progression 8, 12, 18, and 27. Then the successive terms are in the ratio $\frac{3}{2}$, which was known as a *sesquialtera* ratio and, in musical terms, is also known as a perfect fifth. This ratio is also an example of a *superparticularis* ratio—that is, a ratio of the form $\frac{n+1}{n}$, where n is a positive integer.

This reconstruction of Plato's argument provides us with a way to imagine what lies behind a pythagoroean view of the world, but does not necessarily reflect an argument that Plato might have presented himself. The argument he puts forward in the *Timaeus* is much more succinct and somewhat opaque to a modern reader, steeped as it is in the classical Greek philosophical approach to numbers and their relationship with the natural world. The explanation provided by Timaeus hinges on an assertion about planes and solids:

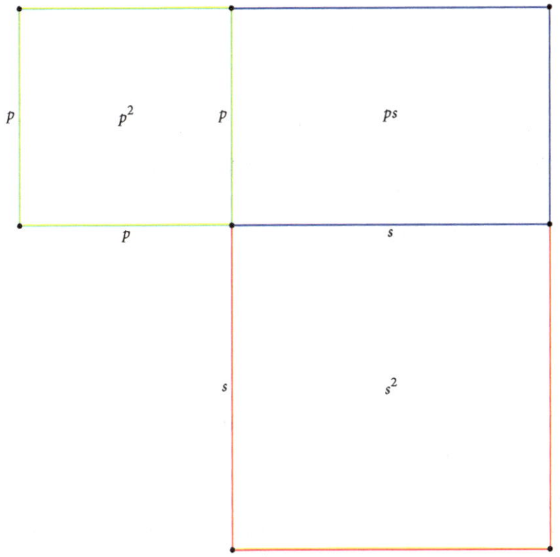

Fig. 3.3: Geometric progression between two squares

> Now if the body of the universe could have been a plane without depth, one middle term would have sufficed to bind together its companions and itself. But in fact the world was to be a solid, and solids must always be conjoined not by one middle term, but by two.[12]

That is, if the universe were two-dimensional, then the contraries of fire and earth could be bridged, continuously, by a single intermediary, as in figure 3.3.[13] But the universe is three-dimensional, and so two intermediaries are required, as we saw in figure 3.2.

This passage has received much attention from antiquity until the present: although clear in reference to geometric squares and cubes (since only the measurement of one dimension at a time can be changed to preserve a continuous progression), it is not literally true in reference to arithmetic squares and cubes. That is, given two integers a and b, we can form a geometric progression between a^2 and b^2 with one term,

$$a^2, ab, b^2,$$

12 Plato, *Plato: Timaeus and Critias*, 32a–b, 29.

13 This diagram is similar to the one found in Proposition 23 of Book VI of Euclid. For a 9th-century rendering, see: Euclid, "In Euclidis elementorum libri," in Manuscript Gr. 190.pt.1 (Vatican: Vatican Library, Digivatlib), f. 97v, accessed February 11, 2026, https://digi.vatlib.it/mss/detail/Vat.gr.190.pt.1; For a modern rendering, see: Euclid, *The Thirteen Books of Euclid's Elements*, vol. 2, 247.

and a geometric progression between a^3 and b^3 with two terms,

$$a^3, a^2b, ab^2, b^3.$$

For example, the progression 4, 6, 9 is an example of the first and 8, 12, 18, 27 of the second. However, if two integers are, at the same time, squares of some integer and cubes of another, such as $64 = 8^2 = 4^3$ and $729 = 27^2 = 9^3$, then either progression may be formed. For this example, we have both

$$64, 216, 729,$$

and

$$64, 144, 324, 729.$$

However, as Thomas Henri Martin[14] and A. E. Taylor[15] have maintained, the apparent difficulty is resolved if we assume both a and b are prime numbers. It seems very reasonable to assume that Plato was considering, in the planar case, integers which were the product of exactly two factors and, in the solid case, integers which were the product of exactly three factors. Proclus presents this same example as a potential objection to Plato's argument, but dismisses it quickly:

> [T]he person who puts this objection uses solids not just in so far as they are solids when he says they are bound by a single middle term. But if he had considered them *qua* solids and *cubes*, he would find that there are two middle terms between them.[16]

In other words, numbers that are both squares and cubes, such as 64 and 729, may be considered either as plane numbers or as solid numbers, and it is clear that Plato is looking at them as representing solids in our three-dimensional world.

3.2 Calcidius

Many commentators and expositors would repeat Plato's formulation of the creation and necessity of the four elements, with little change, over the next centuries. For example, Alcinous, writing in the 2nd century CE,[17] summarized the creation of the elements by the demiurge in language very similar to that of Timaeus:

---

14 Thomas Henri Martin, *Études sur le Timée de Platon* (Paris: Ladrange Libraire-Éditeur, 1841), 339.

15 A. E. Taylor, *A Commentary on Plato's Timaeus* (Oxford: Clarendon Press, 1928), 98.

16 Proclus, *Commentary on Plato's Timaeus*, 3.32, 79.

17 As little is known about Alcinous, dating his work to the 2nd century is conjectural. See: Alcinoos, *Alcinoos: Enseignement des doctrines de Platon*, ed. John Whittaker, trans. Pierre Louis (Paris: Les Belles Lettres, 1990), XII–XIII.

> For he [the demiurge] generated it [the world] out of all of each of the four elements, all of fire, and earth and water, and air, not leaving out any part of potency of any of them, on the consideration that, first of all, what came into being must be corporeal, and so inevitably tangible and visible. But without fire and earth it is not possible for something to be tangible and visible. So, following probable reasoning, he fashioned it out of earth and fire; and since some bond was required to bring both of these together, and the divine bond is that of proportion, whose nature it is to make one both itself and what binds together, and since the world was not a plane figure (for in that case one mean would have sufficed for it), but spherical, it required two means to bring it into harmony. For this reason, then, air and water were placed between fire and earth, following the system of proportion: so, as fire is to air, so air is to water, and this latter in turn to earth, and vice versa.[18]

This formulation is not only a restatement of what Plato wrote in the *Timaeus*, it may be essentially a paraphrase of a 1st-century compilation of the teachings of Plato—namely that of Arius Didymus in his *On the Doctrines of Plato*—although the amount of this earlier treatise borrowed by Alcinous is uncertain.[19]

The platonic philosophical system was known throughout most of the Middle Ages primarily from two partial translations of the *Timaeus*. The translation of Cicero in the 1st century BCE was the first and shorter of these, covering only 27d – 47b.[20] The second translation, by Calcidius, covering 17c – 53c, included a commentary on 31c – 53c and ultimately was more influential in the revival of neoplatonic thought in the later medieval period. However, very little is known of Calcidius himself. He composed his work sometime in the 4th century CE, but whether in the first half or the second half of the century is a matter of some debate. Similarly, there are questions about whether Latin was his first or second language, whether he was a Christian, where he lived, and the identity of Osius, at whose request he undertook the work.[21]

Whatever his background, the commentary of Calcidius clearly shows that, by the 4th century, the three contrary dimensions of fire and earth had come to be identified with three pairs of opposing qualities. Where fire has sharpness (*acumen*),

---

18 Alcinous, *The Handbook of Platonism*, trans. John Dillon (Oxford: Clarendon Press, 1993), 12.2, 21.

19 John Dillon, *The Middle Platonists: 80 B.C. to A.D. 220* (Ithaca, New York: Cornell University Press, 1996), 269; Alcinoos, *Alcinoos: Enseignement des doctrines de Platon*, 109.

20 The Stephanus numbers for the *Timaeus* run from 17a to 92c.

21 For more background on Calcidius, both what is known and what is conjectured, see: Calcidius, *On Plato's Timaeus*, ed. and trans. John Magee (Cambridge, MA: Harvard University Press, 2016), vii–xvii; Calcidius, *Calcidius: Commentaire au Timée de Platon*, ed. Béatrice Bakhouche, trans. Béatrice Bakhouche and Luc Brissson, 2 vols. (Paris: Librairie Philosophique J. Vrin, 2011), 7–13; Calcidius, *Timaeus: A Calcidio translatus commentarioque instructus*, ed. J. H. Waszink (London and Leiden: Aedibus Instituti Warburgiana et E. J Brill, 1975), ix–xvii; Christina Hoenig, *Plato's Timaeus and the Latin Tradition* (Cambridge: Cambridge University Press, 2018), 160–62.

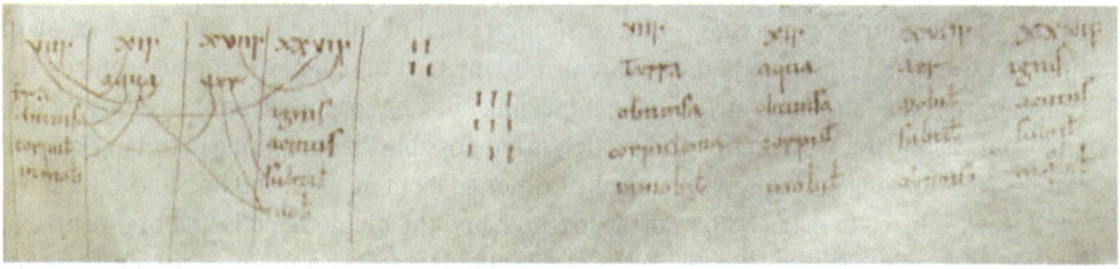

Fig. 3.4: Marginalia in MS Brussels 9625–9626, f. 14r

fineness (*subtilitas*), and mobility (*mobilitas*); earth has compactness (*obtunsitas*), corporeality (*corpulentia*), and immobility (*immobilitas*).[22] Then the first element which is intermediate between fire and earth—namely, air—will have two qualities of fire and one of earth:

> Si enim uicinum igni elementum quod sit et ex quibus conflatum uoluerimus inquirere, sumemus ignis quidem de proximo duas uirtutes, subtilitatem et mobilitatem, unam uero terrae, id est obtunsitatem, et inuenietur genitura secundi elementi quod est subter ignem, id est aeris; est enim aer obtunsus subtilis mobilis.[23]
>
> [For if we wish to inquire into the nature and constituent parts of the element bordering on fire, we will isolate first two powers of fire, its fineness and mobility, and one of earth, i.e., its compactness, and the origins of the element immediately below fire, i.e., air, will be found; for air is compact, fine, and mobile.[24]]

The second intermediate element—namely, water—will then have two qualities of earth, as well as two qualities of air and one of fire:

> Rursumque si eius elementi quod est uicinum terrae, id est aquae, genituram consideremus, sumemus duas quidem terrae uirtutes, id est obtunsitatem et corupulentiam, unam uero ignis, id est motum, et exorietur aquae substantia, quae est corpus obtunsum corpulentum mobile.[25]
>
> [Again, if we reflect on the origins of the element neighboring on earth, i.e., water, we will isolate two powers of earth, i.e., its compactness and corporeality, and one of fire, i.e., its movement, and the substance of water will emerge, that being a body compact, corporeal, and mobile.[26]]

22 Calcidius, *Timaeus: A Calcidio translatus*, XXI, 72; Calcidius, *On Plato's Timaeus*, 21, 150–53.
23 Calcidius, *Timaeus: A Calcidio translatus*, XXII, 72.
24 Calcidius, *On Plato's Timaeus*, 22, 153.
25 Calcidius, *Timaeus: A Calcidio translatus*, XXII, 72.
26 Calcidius, *On Plato's Timaeus*, 22, 153.

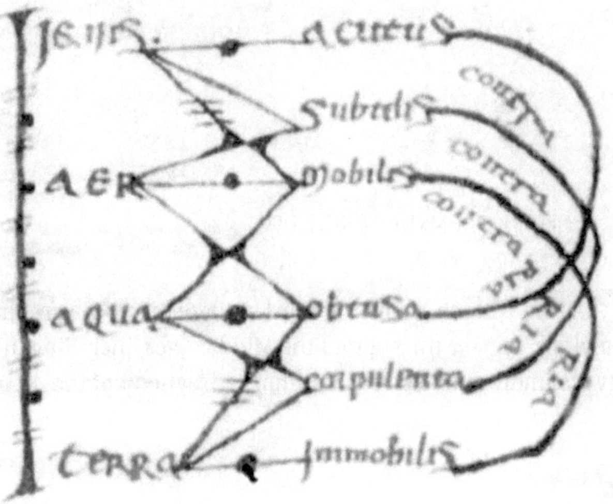

Fig. 3.5: Diagram of the elements in BNF MS Latin 6281, f. 19r

Calcidius illustrates his argument with figures similar to those in figures 3.3 and 3.2, although he labels the vertices rather than the lengths of the sides. In essence, he has identified the three dimensions of space with three pairs of contrasting qualities: *acumen/obtunsitas*, *subtilitas/corpulentia*, and *mobilitas/immobilitas*. To move from fire to air to water to earth, we hold two of these dimensions fixed while changing the remaining one as we move from one element to the next. Geometrically, this is seen in figure 3.2 as stretching one side of a box in each step, as we move from the cube representing earth to the cube representing fire.

Calcidius does not introduce numerical values, and so never considers the particular representative geometric progression 8, 12, 18, 27. However, this did not preclude later scribes adding annotations in this regard in the margins of copies of the text. For example, in the 10th-century manuscript MS Brussels 9625–9626, folio 14r, we find the illustration shown in figure 3.4 in the bottom margin.[27] Here we see a table relating the elements earth, water, air, and fire with the numerical sequence 8, 12, 18, and 27, along with the list of associated properties. And, on the left side of the margin, we see a rudimentary form of a diagram illustrating the relationships between the elements and their properties. Similar diagrams are found in the 12th-century manuscripts BNF MS Latin 6281 and BNF MS Latin 6282, as seen in

27 "Timaeus Platonis," in MS Brussels 9625–26 (KBR: Bibliothèque royale de Belgique), accessed March 2, 2026, https://uurl.kbr.be/bib/17367397.

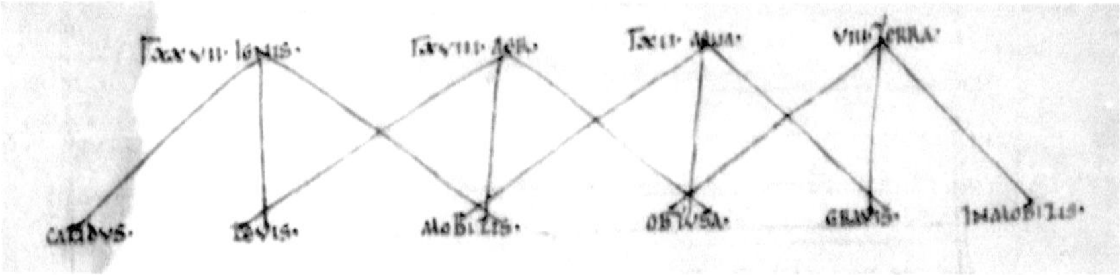

Fig. 3.6: Diagram of the elements in BNF MS Latin 6282, f. 110r

figures 3.5 and 3.6.[28] As we will see below, versions of these diagrams, frequently in more elaborate drawings, appear throughout the Middle Ages, including in AM 736 III, the 16th-century Icelandic manuscript containing a fragment of the *Algorismus*.

3.3 Macrobius

Another influential thread contributing to the spread of neoplatonic philosophy in the Middle Ages began with Macrobius Ambrosius Theodosius. Commonly referred to as Macrobius Theodosius or just Theodosius in the 4th and 5th centuries, but simply as Macrobius throughout most of the medieval period, what little is known about his life is open to numerous interpetations and speculations. He composed his works sometime in the last quarter of the 4th century or the first quarter of the 5th. He says of himself that "nos sub alio ortos caelo,"[29] and so is presumed to have been born at some distance from Rome, perhaps in North Africa, but there is no consensus even on whether his native language was Greek or Latin. Manuscripts attest that he was "vir clarissimus et inlustris," a title indicative of someone holding one of the hightest offices in the Empire. Holding such an office may indicate that he was a Christian, but that is not evident from his writings and is open to some debate.[30]

28 "*Chalcidii* interpretatio et commentarius in Timaeum Platonis," in BNF Latin 6281 (Bibliothèque nationale de France), accessed December 9, 2025, https://archivesetmanuscrits.bnf.fr/ark:/12148/cc65354z; "Timaeus *Platonis*, de graeco in latinum, petente Josio, sive potiùs Osio, Cordubensi Episcopo, à Chalcidio, viro claro translatus, et miro ingenio commentatus et elucidatus," in BNF Latin 6282 (Bibliothèque nationale de France), accessed December 9, 2025, https://archivesetmanuscrits.bnf.fr/ark:/12148/cc653556.

29 Macrobius, *Saturnaliorum*, ed. Ludwig von Jan (Quedlinburg and Leipzig: Godofredius Bassius, 1852), Praefatio 11, 7.

30 For more on the speculations on the life of Macrobius, see: Macrobius, *Commentary on the Dream of Scipio*, 3–9; Alan Cameron, "The Date and Identity of Macrobius," *The Journal of Roman Studies* 56 (1966): 25–38; Macrobius, *Commentarii in Ciceronis somnium Scipionis*, II–VII.

Although not particularly original nor of great interest to a modern scholar of neoplatonic thought, Macrobius's *Commentarii in somnium Scipionis* was one of the most influential works in the development of neoplatonic ideas in the succeeding centuries.[31] Within this treatise, which is, overall, a commentary on a section of the last book of Cicero's *De re publica*, Macrobius repeats the account of the creation of the elements from the *Timaeus*. Quoting Plato, he writes:

> Unde omne mundi corpus de igne et terra instituere fabricator incipiens vidit duo convenire sine medio colligante non posse, et hoc esse optimum vinculum quod et se pariter et a se lig
anda devinciat: unam vero interiectionem tunc solum posse sufficere, cum superficies sine altitudine vincienda est: at, ubi artanda vinculis est alta dimensio, nodum nisi gemina interiectione non conecti. Inde aerem et aquam inter ignem terramque contexuit : et ita per omnia una et sibi conveniens iugabilis conpetentia cucurrit elementorum diversitatem ipsa differentiarum aequalitate consocians.[32]

> [Hence the Creator, endeavoring to compound the body of the universe from fire and earth, saw that the two could not be combined without some mean to cement them, and that the mean which was best would bind itself as well as the extremes. He also saw that one mean would be sufficient only to bind surfaces without altitude, but that when a third dimension had to be included in the union, two means were necessary to bind the whole. Accordingly, he wove air and water into fire and earth, and thus a mutual attraction ran through the universe, linking together unlike elements by the similarities underlying their differences.[33]]

However, he does not analyze the progression between the elements in terms of the three contrary qualities described by Calcidius. Rather, he uses the two contraries dry/wet and cold/hot:

> Ita enim elementa inter se diversissima opifex tamen deus ordinis oportunitate conexuit, ut facile iungerentur: nam, cum binae essent in singulis qualitates, talem unicuique de duabus alteram dedit, ut in eo cui haereret cognatam sibi et similem repperiret. Terra est sicca et frigida, aqua vero frigida et humecta est: haec duo elementa, licet sibi per siccum humectumque contraria sint, per frigidum tamen commune iunguntur. Aer humectus et calidus est: et, cum aquae frigidae contrarius sit calore, conciliatione tamen socii copulatur humoris. Super hunc ignis, cum sit calidus et siccus , humorem quidem aeris respuit siccitate, sed conectitur per societatem caloris. Et ita fit, ut singula quaeque elementorum duo sibi hinc inde vicina singulis qualitatibus velut quibusdam amplectantur ulnis: aqua terram frigore, aerem sibi nectit humore: aer aquae bumecto simili, et igni calore sociatur: ignis aeri miscetur ut calido, terrae iungitur siccitate: terra ignem sicco patitur, aquam frigore non respuit.[34]

31 Macrobius, *Commentary on the Dream of Scipio*, 10.
32 Macrobius, *Commentarii in Ciceronis somnium Scipionis*, I.VI.30–31, 43.
33 Macrobius, *Commentary on the Dream of Scipio*, I.VI.30–31, 105–6.
34 Macrobius, *Commentarii in Ciceronis somnium Scipionis*, I.VI.25–27, 42.

> [For thus, in spite of the utter diversity of these elements, the Creator harmonized them so skillfully that they could be readily united. To each of them he gave two qualities, one of which was of such sort that each element would find this quality related and similar to itself in the element to which it adhered. Earth is dry and cold, and water cold and moist; but although these two elements are opposed, the dry to the wet, they have a common bond in their coldness. Air is moist and warm and, although opposed to water, the cold to the warm, nevertheless has the common bond of moisture. Moreover, fire, being hot and dry, spurns the moisture of the air, but yet adheres to it because of the warmth in both. And so it happens that each one of the elements appears to embrace the two elements bordering on each side of it by single qualities: water binds earth to itself by coldness, and air by moisture; air is allied to water by its moisture, and to fire by warmth; fire mingles with air because of its heat, and with earth because of its dryness; earth is compatible with fire because of its dryness, and with water because of its coldness.[35]]

In his commentary on the *Timaeus*, Proclus attributes this characterization of the elements to Ocellus Lucanus, an early pythagorean philosopher whose name, along with the names of his brother and sister, appears in a list of pythagorean philosophers compiled by Iamblichus.[36] However, the text in question, *On the Nature of the Universe*, once attributed to Ocellus, has since been shown to be a forgery. The true author, now referred to as Pseudo-Ocellus, remains unknown. And, contrary to the supposition of Proclus that the work dated from before Plato, it was most likely written between 200 and 150 BCE.[37] Pseudo-Ocellus gives a short description of the elements in terms of contraries within a discussion of their creation and mutation:

> Fire and earth, therefore, are the extremities of the elements, while water and air are the media, for they have a mixed corporeal nature. Nor is it possible that there could be only one of the extremes, a contrary thereto being necessary. Nor could there only be two, for it is necessary to have a medium, as media oppose extremes.
> Fire therefore is hot and dry, but air is hot and moist; water is moist and cold, and earth is cold and dry. Hence heat is common to air and fire; cold is common to water and earth; dryness to earth and fire, and moisture to water and air.[38]

This characterization of the elements is not original to Pseudo-Ocellus and is not part of the platonic tradition. In fact, the discussions in *On the Nature of the Universe* appear to be in line with the Peripatetic school and derive, ultimately, from Aristotle

35 Macrobius, *Commentary on the Dream of Scipio*, I.VI.25–27, 105.

36 Iamblichus, *De vita pythagorica*, ed. Ludovicus Deubner (Stuttgart: B. G. Teubner, 1975), 36, 146; Kenneth Sylvan Guthrie, ed. and trans., "On the Nature of the Universe," in *An Anthology of Ancient Writings Which Relate to Pythagoras and Pythagoroean Philosophy* (Grand Rapids, MI: Phanes Press, 1987), 203.

37 Proclus, *Commentary on Plato's Timaeus*, 84; Holger Thesleff, *An Introduction to the Pythagorean Writings of the Hellenistic Period* (Turku, Finland: Åbo Akademi, 1961), 71, 102, 115.

38 Guthrie, "On the Nature of the Universe," 207.

and his student Aristoxenos. In particular, this passage resembles what Aristotle wrote in his work *On Generation and Corruption*:

> The elements are four, and any four terms can be combined in six couples. Contraries, however, refuse to be coupled; for it is impossible for the same thing to be hot and cold, or moist and dry. Hence it is evident that the couplings of the elements will be four: hot with dry and moist with hot, and again cold with dry and cold with moist. And these four couples have attached themselves to the *apparently* simple bodies (Fire, Air, Water, and Earth) in a manner consonant with theory. For Fire is hot and dry, whereas Air is hot and moist (Air being a sort of vapour); and Water is cold and moist, while Earth is cold and dry.[39]

Proclus rejected this description of the progression of the elements, arguing, for example, that this characterization of the elements makes earth and fire more alike than either earth and air or water and fire. Hence, in this system, it is not the most extreme elements, earth and fire, that are the most contrary, but rather the most contrary elements are fire and water as well as air and earth. Moreover, he argues, proximate elements disagree to the same extent that they agree, whereas we would expect nearby elements to agree more than they disagree.[40]

We would expect that any diagrams that later scribes might add to the text of the *Commentarii in somnium Scipionis* would illustrate the elements in terms of the Aristotelian contraries. This is indeed what is found in, for example, the 12th-century manuscript NKS 218, folio 3r, in the Kongelige Bibliotek in Copenhagen, as we see in figure 3.7.[41] Yet we also find that some scribes made the identification with the neoplatonic contraries described by Calcidius and, in particular, with the sequence 8, 12, 18, 27. For example, the manuscript Pal. lat. 1341, folio 94v, in the Biblioteca Vaticana, written sometime in the second half of the 10th century, contains the *Commentarii in somnium Scipionis*, as well as the *Somnium Scipionis* itself,

39 Aristotle, "On Generation and Corruption," in *The Complete Works of Aristotle: The Revised Oxford Translation*, ed. Jonathan Barnes, trans. H. H. Joachim, vol. 1 (Princeton, New Jersey: Princeton University Press, 1984), 330a30–330b5, 540.

40 Proclus, *Commentary on Plato's Timaeus*, 3.38, 84–85; Thomas Piketty has done a similar analysis of voting blocs in Western Europe and the United States. Using the contraries internationalist/nationalist and egalitarian/inegalitarian, he identifies the left with internationalist and egalitarian, the center-left with internationalist and inegalitarian, the center-right with nationalist and inegalitarian, and the right with nationalist and egalitarian. Whereas the resulting instability works against this type of model for the four elements, it accurately reflects current political instabilities. See: Thomas Piketty, *Capital and Ideology*, trans. Arthur Goldhammer (Cambridge, MA: The Belknap Press of Harvard University Press, 2020).

41 NKS 218 4to: Macrobius, Commentarii in somnium Scipionis (Det Kongelige Bibliotek), 3r, accessed December 9, 2025, https://permalink.kb.dk/permalink/2006/manus/33/eng/.

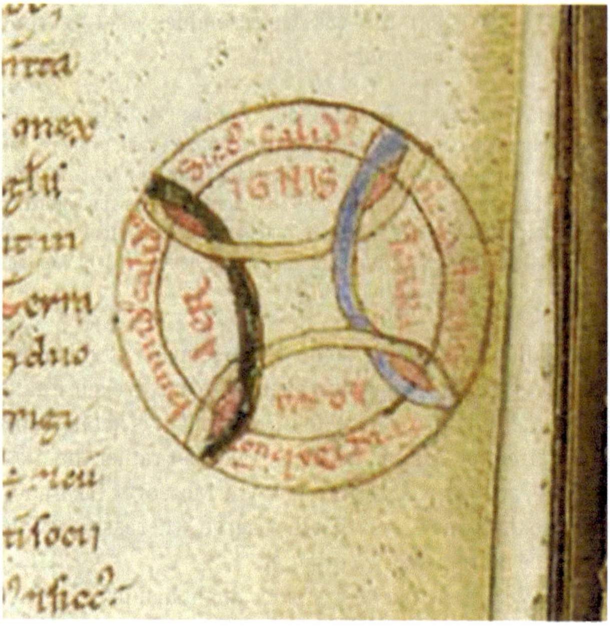

Fig. 3.7: Diagram of the elements with Aristotelian contraries, NKS 218, f. 3r

followed by the diagram of the elements reproduced in figure 3.8.[42] The diagram is also interesting as it reverses the order of the identification of the elements with the geometric sequence 8, 12, 18, and 27 from what appears in figures 3.4 and 3.6. Here 8 represents fire, 12 represents air, 18 represents water, and 27 represents earth. Of course, only the proportionality between the numbers, not the numbers themselves, are relevant to the meaning of the neoplatonic framework. We will see a similar reversal in figure 3.12 below.

3.4 Proclus

In contrast to Calcidius and Macrobius, the life of the 5th-century neoplatonist philosopher Proclus is well documented. He was born in Byzantium, lived there as well as in Xanthus and Alexandria as a youth, and began studies at the Academy in Athens under Syrianus when he was nineteen. He would later succeed Syrianus as

42 Pal. lat. 1341 (Biblioteca Apostolica Vaticana, Digivatlib), 94v, accessed February 23, 2026, https://digi.vatlib.it/view/bav_pal_lat_1341/0196.

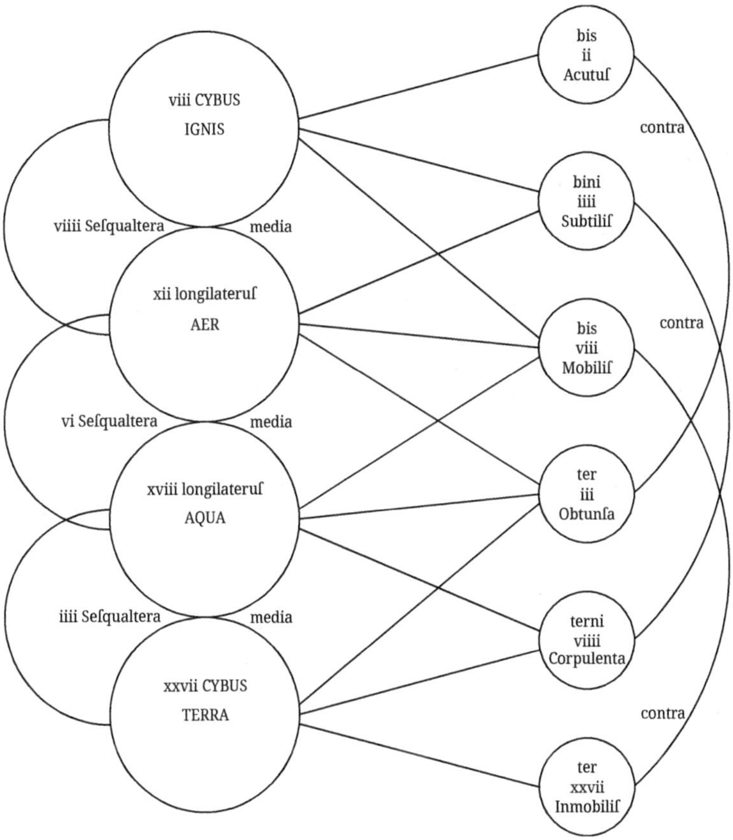

Fig. 3.8: Reproduction of a diagram of the elements in Biblioteca Vaticana, Pal. lat. 1341, f. 94v

head of the Academy while still in his twenties, a position he would maintain until his death in 485 CE.[43]

In his commentary on the *Timaeus*, Proclus provides the clearest explanation of what had become the neoplatonic description of the relationships between the four elements. Having argued, as we saw above, that two contraries are not sufficient to describe the proportionalities between the elements, he summarizes the properties of the elements in a table, as shown in table 3.1, using the same contrary characteristics as Calcidius.[44] He then explains:

---

43 Proclus, *Commentary on Plato's Timaeus*, ed. and trans. Harold Tarrant, vol. I (Cambridge: Cambridge University Press, 2007), 2–3.

44 Proclus, *Commentary on Plato's Timaeus*, 3.39, 87.

Tab. 3.1: Properties of the elements

Element	Quality		
Fire	tenuousness	sharpness	easy mobility
Air	tenuousness	bluntness	easy mobility
Water	density	bluntness	easy mobility
Earth	density	bluntness	immobility

> Therefore, since fire and earth are similar bodies and similar solids, two proportional middle terms will fall between them and each of these middles [will be a solid] having two sides from the extreme term closer to it and one side from the remaining one. So since fire will have three physical sides – that is, the three powers: tenuousness, sharpness and easy mobility – then taking away the middle term, sharpness, and substituting bluntness will make air. Air has two sides from fire, but a single one from earth[45] or [we may also say] it has two powers from fire, but a single power from earth. So it also turns out that it is fitting that air has more association with what is near it rather than what is the third from it in terms of distance or separation.
> Once again, since earth has three physical sides which are contraries to the powers of fire – density, bluntness and immobility – then by taking away immobility and replacing it with ease of mobility, water will be produced. Since water has density, bluntness and ease of mobility, it has two sides or powers in common with earth, but takes only one from fire. So surely in this way these middle terms and the elements will be innately conjoined, communing through two powers, requiring two means for the sake of likeness, but differing in one. And in this fashion each one will always be conjoined to a greater degree with the one that is closer to it than it is with the one further away from it. And a single cosmos will result through all the elements and a single harmonic order from the dominance of proportion.[46]

Going further than Calcidius, Proclus explicitly connects the proportionalities between the elements with the geometric progression between two cubes, illustrating it with the sequence connecting 8 and 27:

> It is like this too in the case of the middle terms between two cubes like 8 and 27. 12 (which is the middle term that is alongside 8) will have two sides from 8 but only one from 27, for 12 is $2 \times 2 \times 3$. But it is the other way around with 18. Since it is $3 \times 3 \times 2$, it has two sides from 27 (the two 3s) but one side from 8 – that is, the 2. Plato's beliefs about the physical nature of the elements of the universe are thus in concord with mathematics.[47]

45 Baltzy's translation erroneously has "air" instead of "earth" (γῆς). See: Proclus, *In Platonis Timaeum commentaria*, ed. Ernst Diehl (Leipzig: B. G. Teubner, 1904), 40.27.

46 Proclus, *Commentary on Plato's Timaeus*, 3.40–41, 88–89.

47 Proclus, *Commentary on Plato's Timaeus*, 3.41, 89.

Although, as noted above, Proclus warns about conflating physical figures with mathematical numbers, at the same time he recognizes the insights we gain from considering analogies between the mathematical and physical points of view:

> After taking up the meaning of these words from a mathematical point of view, it is necessary to be directed to the consideration of their meaning with respect to physical theory. For it is not appropriate either to get bogged down in these mathematical aspects and suspend the rest of the discourse (for the dialogue is about physics) or to ignore the preceding words and examine only that which bears on the sensible realm. Rather it is necessary to bring both aspects together and constantly interweave the mathematical with the physical.[48]

Proclus understands that the role that mathematics plays in the study of nature is both subtle and complex.

In all, Proclus provided the clearest and most comprehensive explanation of the neoplatonic description of the elements. The formulation seen in his commentary on the *Timaeus* will be repeated many times, in slightly varying forms and usually without the broader discussion on the place of mathematics in the study of physical nature, throughout the Middle Ages. However, this work itself appears to have been unknown in Western Europe between the time of Boethius and the late 13th century. In the 11th century, the Greek monk Michael Psellos refers to the commentary of Proclus in his writing, yet in the West the first translation does not appear until William of Moerbeke translated three extracts on prayer in the 13th century. A full translation in Latin would not appear until the Renaissance.[49]

3.5 Arithmetic: Nicomachus and Boethius

Another thread of commentary on Plato's remarks in the *Timaeus* on geometric sequences between squares and cubes, but not on the connection with the elements, may be found in treatises on arithmetic. In particular, such commentaries may be found in the *Introduction to Arithmetic* of Nicomachus of Gerasa and the *De Institutione Arithmetica* of Boethius.

Nicomachus was a neopythagorean mathematician and philosopher. Although highly regarded as a mathematician and philosopher both during his life and long after, very little is known of his actual biography. He was born in Gerasa (modern day Jerash in Jordan), one of the Hellenistic cities in the Levant. His life spanned a

48 Proclus, *Commentary on Plato's Timaeus*, 3.23, 68.

49 Stephen Gersh, "One Thousand Years of Proclus: An Introduction to His Reception," in *Interpreting Proclus: From Antiquity to the Renaissance*, ed. Stephen Gersh (Cambridge: Cambridge University Press, 2014), 26.

period sometime between the mid-points of the 1st and 2nd centuries, but not much more can be said with certainty.[50]

In his discussion of proportions in the second book of his *Introduction to Arithmetic*, Nicomachus addresses Plato's claims about solid and plane numbers in the *Timaeus*. Although not explicitly stated, he must realize that Plato's contention that two intermediate terms are necessary in order to connect two cubes by a geometric sequence is, as discussed above, not always true. To resolve the difficulty, he restricts his discussion to the squares and cubes of consecutive integers:

> It would be most seasonable, now that we have reached this point, to mention a corollary that is of use for a certain Platonic theorem: for plane numbers are bound together always by a single mean, solids by two, in the form of a proportion. For with two consecutive squares only one mean is discovered which preserves the geometric proportion, as antecedent to the smaller and consequent of the greater term, and never more than one. Hence we conceive of two intervals between the mean term and each extreme, in the relation of similar ratios. Again, with two consecutive cubes only two middle terms in proper ratio are found, in accordance with the geometric proportion, never more; hence there are three intervals, one, that between the mean terms compared to one another, and two between the extremes and the means on either side.[51]

For squares, he provides the example

$$4, 6, 9,$$

which he further explains geometrically:

> The reason for this is that the sides of the two squares [namely, 2 and 3], one belonging peculiarly to each, both together produced this very number 6.[52]

The reference is to squares, as we saw above in figure 3.3, with $p = 2$ and $s = 3$. Similarly, for cubes, as we saw above in figure 3.2 (again with $p = 2$ and $s = 3$), Nicomachus explains:

> In cubes, however, for example 8 and 27, no longer one but two mean terms are found, 12 and 18, which put themselves and the terms in the same ratio as that which the differences bear to one another; and the reason of this is that the two mean terms are the products of the sides commingled, 2 times 2 times 3 and 3 times 3 times 2.[53]

50 For what details are know of the life of Nicomachus, see: Nicomachus, *Introduction to Arithmetic*, ed. Martin Luther D'Ooge (New York: The Macmillan Company, 1926), 71–78; Dillon, *The Middle Platonists*, 352–53.

51 Nicomachus, *Introduction to Arithmetic*, II.XXIV.6, 272–73.

52 Nicomachus, *Introduction to Arithmetic*, II.XXIV.9, 273.

53 Nicomachus, *Introduction to Arithmetic*, II.XXIV.9, 273–74.

Although Nicomachus clearly has the *Timaeus* in mind here, he restricts his comments to the topic of proportions between numbers and makes no mention of any connection to proportional relationships between the elements. It is possible that he made this connection elsewhere, perhaps in his work the *Theology of Arithmetic*, which is now lost except for some excerpts preserved by later authors.[54] Given his neopythagorean philosophical outlook, this would not be unexpected.

The *Introduction to Arithmetic* of Nicomachus became central to Western European education and thought through a 6th-century translation into Latin by Boethius. Anicius Manlius Severinus Boethius was born into an aristocratic Roman family sometime around the year 480 CE and, while still a youth, was adopted and later married into the even more eminent household of Quintus Aurelius Memmius Symmachus. He availed himself of the educational oppportunities his position in society presented to him. In particular, he learned Greek and much of what he wrote reflects his extensive knowledge of the neoplatonic works of the Athenian and Alexandrian neoplatonists, to such an extent that some have hypothesized that he studied in Alexandria at some point during his formative years.[55] Often seen as a bridge between philosophers of Late Antiquity and those of the early years of medieval scholasticism, the extent to which Boethius could have contributed to the dissemination of Greek authors in Western Europe was diminished by his execution around the year 525 CE.[56]

The *De institutione arithmetica* is one of the earliest works of Boethius, composed between the years 500 and 510 CE when he was in his twenties.[57] Presenting the result of his labor to Symmachus, whom Boethius asks to evaluate his work while praising Symmachus for his knowledge of both Greek and Latin, Boethius declares at the outset that his intention was never to produce a translation tightly bound to the words of Nicomachus, but rather has abridged the original text when he felt Nicomachus went on for too long, and provided additional explanation when he thought the original was too quick for an adequate understanding.[58] Nicomachus's treatment of the geometric progression between plane and solid numbers

---

54 Dillon, *The Middle Platonists*, 352–53.

55 Henry Chadwick, *The Consolations of Music, Logic, Theology, and Philosophy* (Oxford: Clarendon Press, 1981), 20.

56 Chadwick, *The Consolations of Music*, 54–55; For more on the life of Boethius, see: Chadwick, *The Consolations of Music*, 1–68; Noel Harold Kaylor Jr., "Introduction: The Times, Life, and Work of Boethius," in *A Companion to Boethius in the Middle Ages*, ed. Noel Harold Kaylor Jr. and Philip Edward Phillips (Leiden: Brill, 2012), 1–46; Boethius, *Boèce: Institution arithmétique*, ed. Jean-Yves Guillaumin (Paris: Les Belles Lettres, 1995), VII–XIV.

57 Kaylor, "The Times, Life, and Work of Boethius," 23.

58 Boethius, *Boèce: Institution arithmétique*, Praef., 2–3; Boethius, *Boethian Number Theory: A Translation of the De Insitutione Arithmetica*, trans. Michael Masi (Amsterdam: Rodopi, 1983), 67.

falls into the latter category. In his expanded presentation, Boethius not only provides additional details and explanation, but also begins by clearly identifying the *Timaeus* as the source of the proposition in question:

> Post haec igitur tempus est ut expediamus nunc quiddam nimis utile in platonica quadam disputatione, quae in Timaei cosmopoeia haud facili cuisqam uel penetrabili ratione uersatur. Omnes enim planae figurae, quae nulla altitudine crescunt, una tantum medietate geometrica continuantur; alia quae iungat non potest inveniri; unde duo tantum in his interualla sunt constituta, a primo scilicet ad medium et a medio ad tertium. Si uero fuerint cybi, duas tantum habebunt medietates, ubi tertia inueniri non poterit secundum geometricam scilicet proportionem; unde formae solidae tria interualla dicuntur habere. Est enim unum interuallum a primo ad secundum et a secundo ad tertium et a tertio ad quartum, quae est scilicet postrema distantia. Recte igitur et planae figurae duobus interuallis et solidae tribus contineri dicuntur.[59]

> [After these matters, it is now time we relate something very useful in Platonic disputation which is treated in that all encompassing work, the *Timaeus*, and which is by no means easy for anyone unless he is trained in a very discerning reason. All plane figures which do not grow by altitude are extended only by one kind of geometric medial proportion; no other may be found to keep it together. So, two relationships are constituted in such an interval, that is from the first term to the middle, from the middle to the the third term. Now if they were cubes, they would have two middle numbers where a third relation would not be found according just to geometric proportion; hence solid forms are said to have three intervals. There is one interval from the first to the second, from the second to the third, from the third to the fourth, which is the most distant. Rightly therefore are plane figures said to be contained with two intervals and solid to be contained with three.[60]]

Here Boethius does not restrict himself to the squares or cubes of consecutive integers, as did Nicomachus, but emphasizes the geometric construction of intermediaries between squares and cubes.

As Nicomachus did, Boethius illustrates the two cases with the progressions

$$4, 6, 9$$

for squares and

$$8, 12, 18, 27$$

for cubes. For example, in the case of cubes, he explains the insertion of the intermediaries in detail:

> Si autem cybi sunt, ut VIII et XXVII, duae tantum inter hos eadem proportione medietates constitui queunt, XII scilicet et XVIII. Namque XII ad VIII et XVIII ad XII et XXVII ad XVIII sesqualtera tantum proportione iunguntur. In his quoque eadem laterum ratio est. Namque ex uno

59 Boethius, *Boèce: Institution arithmétique*, II.46, 153–54.

60 Boethius, *Boethian Number Theory*, II.46, 173.

> cybo qui propinquior est una medietas duo latera colligit, ex alternatim uero posito unum. In alia quoque medietate idem est. Ponantur enim duo cybi et in medio eorum duae medietates, quas superius diximus: VIII, XII, XVIII, XXVII. Octonarii igitur latus est binarius: bis enim bini bis octonarium referunt; ternarius uero XXVII cybi latus est; ter enim tres ter XXVII restituunt. Medietas igitur quae iuxta octonarium est, id est XII, mutuatur duo latera ex propinquo sibi octonario et aliud unum latus ex altrinsecus posito XXVII cybo. Bis enim bini ter XII pandunt. Et XVIII eadem ratione duo latera a propinquo sibi XXVII cybo colligit et unum ab altrinsecus posito octonario. Tres enim ter bis XVIII concludunt.[61]
>
> [If these are cubes, as 8 and 27, two such numbers are constituted with the same proportion of the median between them, namely 12 and 18. Now 12 to 8 and 18 to 12 and 27 to 18 are joined by a sesquialter proportion only. In these also there is the same ratio of sides. For from one cube, which is the closer, one median reaches two sides, each one by an alternately posed term. It is also the same in the other medial proportion. Let us put down to cubic numbers and between them the two median terms, which we have named above: 8, 12, 18, 27. The side of eight is binary; twice two gives eight. The side of the cubic number 27 is ternary; three times three thrice gives 27. Thus, the median number which is the next to eight is 12. It is affected by the two sides next to it, from eight on one side and on the other by the cubic number 27. Two twos three time times give 12. Also 18 by the same logic is affected by the two numbers next to it, on the one side by the cubic number 27, on the other side by twice 8. Three threes twice gives 18.[62]]

With this elaboration on the text of Nicomachus, Boethius echoes what Proclus, with whose writings he was well acquainted,[63] had written before and the same will be echoed by numerous authors after Boethius. Yet he does not connect the sequence 8, 12, 18, 27 with the four elements, although his reference to the *Timaeus* shows that he clearly has the connection in mind.

The *De institutione arithmetica* of Boethius would become a standard text in the quadrivium curriculum for centuries after his death.[64] And, as we will see later, even though Boethius did not make reference to the neoplatonic theory of the elements in this text, later scribes and commentators would make the connection. Moreover, as we will also discuss later, his most famous work, *De consolatione philosophiae*, written shortly before his death, directly addresses the neoplatonic theories.

---

61 Boethius, *Boèce: Institution arithmétique*, II.XLVI, 154–55.

62 Boethius, *Boethian Number Theory*, II.46, 173–74.

63 Chadwick, *The Consolations of Music*, 20.

64 Ann E. Moyer, "The *Quadrivium* and the Decline of Boethian Influence," in *A Companion to Boethius in the Middle Ages*, ed. Noel Harold Kaylor Jr. and Philip Edward Phillips (Leiden: Brill, 2012), 479–517.

3.6 Isidore, Bede, and the Nature of Things

3.6.1 Isidore of Seville

Not surprisingly, works on natural philosophy provide further insight into the influence of the neoplatonic conception of the elements on medieval thought. One prominent example of such studies is provided by the *De natura rerum* of Isidore of Seville. Isidore was born around the year 560 CE and grew up in the province of Baetica in Spain. Although Isidore's father died when he was still a youth, Isidore's education was overseen by his oldest brother, Leander, possibly at the cathedral school or at the monastery school where Leander was a monk. Leander would later become bishop of Seville, with Isidore succeeding him in that position around the year 600 CE.[65]

Isidore wrote *De natura rerum* sometime between 612 and 615 CE, possibly in response to a number of lunar eclipses in in the years 611 and 612, as well as a solar eclipse in the latter.[66] As he states in the preface, his goal in composing this treatise was to allay superstitious notions about the natural world with reasoned thought:

> Quae omnia, secundum quod a ueteribus uiris ac maxime sicut in litteris catholicorum uirorum scirpta sunt, proferentes, breui tabella notamus. Neque enim earum rerum naturam noscere superstitiosae scientiae est, si tantum sana sobriaque doctrina considerentur.[67]
>
> [I have addressed all these matters in a brief document, presenting them just as they were formulated by the scholars of antiquity and especially in the works of catholic authors. For to know the nature of these things is not superstitious knowledge, as long as they are investigated in accordance with sound and sensible teaching.[68]]

When he explains the nature of the four elements, he provides a neoplatonic description similar to that of Calcidius and Proclus, but with one change: whereas, to be consistent with Calcidius and Proclus, he should describe air as *mobilis* (mobile), *obtunsus* (blunt), and *tenuis* (thin), he instead uses the terms *mobilis* (mobile), *acutus* (sharp), and *crassus* (thick):

> Partes mundi quattuor sunt: ignis, aer, aqua, terra. Quarum haec est natura: ignis tenuis, acutus et mobilis; aer mobilis, acutus et crassus; aqua crassa, obtunsa et mobilis; terra crassa, ob-

65 For more on the life of Isidore of Seville, see: Isidore of Seville, *On the Nature of Things*, ed. and trans. Calvin B. Kendall and Faith Wallis (Liverpool: Liverpool University Press, 2016), 3–7.
66 Isidore de Seville, *Traité de la nature*, ed. and trans. Jacques Fontaine (Bordeaux: Féret et Fils, Éditeurs, 1960), 1–6.
67 Isidore de Seville, *Traité de la nature*, Praef. 2, 167.
68 Isidore of Seville, *On the Nature of Things*, Preface 2, 107.

tunsa, inmobilis. Quae etiam sibi ita conmiscuntur. Terra quidem crassa, obtunsa et inmobilis, cum aquae crassitudine et obtunsitate conligatur. Deinde aqua aeri crassitudine et mobilitate coniungitur. Rursus aer igni communione acuti et mobilis conligatur. Terra autem et ignis a se separantur, sed a duobus mediis aqua et aere iunguntur.[69]

[There are four parts of the world: fire, air, water, and earth. This is their nature: fire is thin, sharp, and mobile; air is mobile, sharp, and thick; water is thick, blunt, and mobile; and earth is thick, blunt, and immobile. These are also mixed together as follows. Earth, which is indeed thick, blunt, and immobile, combines with the thickness and bluntness of water. Then water is united with the thickness and mobility of air. In turn air combines with fire by reason of their common properties of sharpness and mobility. However, earth and fire are separated from each other, but they are connected by the two middle parts, water and air.[70]]

In addition to the change in the properties of air, Isidore adds a diagram which may be original to him, as he indicates in the way he introduces it:

Haec itaque ne confusa minus colligantur, subiecta expressi pictura.[71]

[Therefore, lest these bewildering matters not be understood, I have represented them by the adjoining picture.][72]

The figure that follows has many variations in the manuscripts, with one of the earliest (from the 8th-century manuscript BNF MS Latin 6413, folio 4v) shown in figure 3.9.[73] A later version, from the early 11th-century manuscript Staatsbibliothek Bamberg Msc.Nat.1, folio 15r, is shown in figure 3.10.[74] The figures are clearly intended to be seen three-dimensionally, as indicated by the inscription “haec figura solida est secundum geometricam rationem” [“THIS FIGURE is solid according to geometrical proportion”[75]] on figure 3.9.[76] Moreover, “secundum geometricum rationem” is likley to be a reference to the geometric progression 8, 12, 18, 27, although the numerical sequence does not appear either in the diagram or in the text.

69 Isidore de Seville, *Traité de la nature*, XI.1, 213.

70 Isidore of Seville, *On the Nature of Things*, 11.1, 130.

71 Isidore de Seville, *Traité de la nature*, XI.1, 213.

72 Isidore of Seville, *On the Nature of Things*, 11.1, 131.

73 “*Isidori Junioris* liber de natura rerum: initium desideratur,” in BNF Latin 6413 (Bibliothèque nationale de France), 4v, accessed December 10, 2025, https://archivesetmanuscrits.bnf.fr/ark:/12148/cc65498x.

74 “De natura rerum,” in Staatsbibliothek Bamberg Msc.Nat.1 (Staatsbibliothek Bamberg), 15r. Photograph: Gerald Raab, accessed December 10, 2025, https://nbn-resolving.org/urn:nbn:de:bvb:22-dtl-0000025375.

75 Isidore of Seville, *On the Nature of Things*, 131.

76 Figure 3.10 has instead “haec figuru solida est iuxta geometricam rationionem,” with a similar meaning.

Fig. 3.9: Cube of the elements in BNF MS Latin 6413, f. 4v

John Murdoch labels this figure the *cybus elementorum* and suggests that no medieval scribe, at least in the extant manuscripts of the *De natura rerum*, accurately reproduced what Isidore had intended.[77] He conjectures that what Isidore truly intended was along the lines of the diagram shown in figure 3.11. Here the three sets of contrary properties—that is, *acutus/obtusus*, *tenuis/crassus*, and *mobilis/inmobilis*—are represented by opposite sides of the cube, with the four elements corresponding to vertices at the intersection of three sides with the associated properties. The remaining four vertices represent combinations of properties which are considered impossible.[78]

While intriguing, others consider Murdoch's reconstruction doubtful. Michael Gorman notes that Isidore does not use the word *cybus* anywhere else in the text and doubts that this rendering could have been the archetype for the figure in the *De natura rerum*.[79] Bruce Eastwood considers the reconstruction perhaps satisfying

77 John E. Murdoch, *Album of Science: Antiquity and the Middle Ages* (New York: Charles Scribner's Sons, 1984), 280.

78 For example, the lower vertex on the right would represent something with the impossible configuration of being *tenuis* and *acutus*, while at the same time *inmobilis*.

79 Michael M. Gorman, "The Diagrams in the Oldest Manuscripts of Isidore's *De natura rerum*," *Studi Medievali* 42, no. 2 (2001): 531–32.

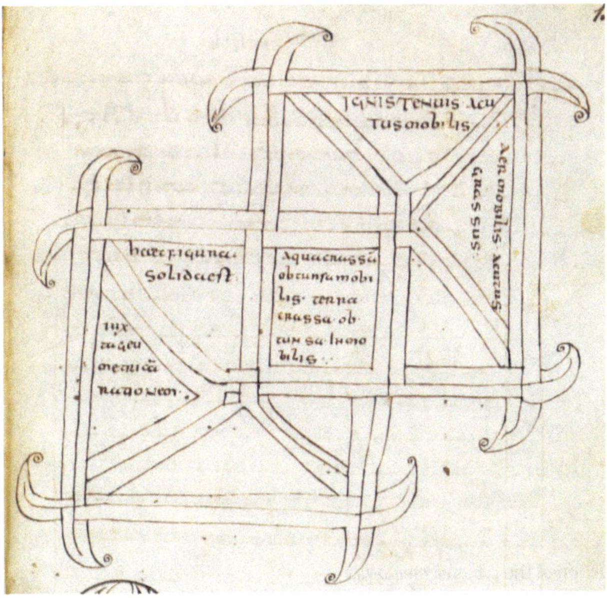

Fig. 3.10: Cube of the elements in Msc. Nat. 1, f. 15r

from a modern perspective, but doubts that it represents Isidore's original figure, and has his own thoughts on what may have been the original intention.[80] On the other hand, Marek Otisk finds Murdoch's suggestion interesting, and offers his own reconstruction.[81] And Barbara Obrist notes that, although Murdoch's explanation offers one possible path for interpretation, it is unlikely that we know enough of the context to provide an accurate reconstruction.[82]

Jacques Fontaine noted that this is the only figure that Isidore introduces in the text using the first person: *subiecta expressi pictura.*[83] This would appear to indicate that the diagram originates with Isidore himself, and is intended to provide what he says it does—that is, a picture to clearly present the relationship between the elements and the three pairs of contrary properties. In that context, Murdoch's re-

80 Bruce S. Eastwood, "The Diagram of the Four Elements in the Oldest Manuscripts of Isidore's *De natura rerum,*" *Studi Medievali* 42, no. 2 (2001): 547–48.

81 Marek Otisk, "The Four Elements and Their Characteristics According to the Schema in the Early Medieval Anonymous Fragment *De Quattuor Elementis,*" *Cosmos and History: The Journal of Natural and Social Philosophy* 17, no. 3 (2021): 469, 471.

82 Barbara Obrist, *La cosmologie médiévale. Textes et images I.* Les fondements antiques (Florence: Sismel Edizioni del Galluzzo, 2004), 276.

83 Isidore de Seville, *Traité de la nature*, 17.

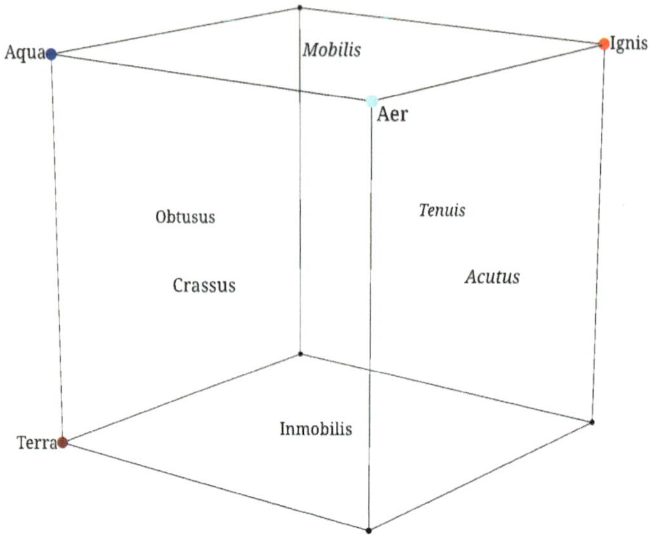

Fig. 3.11: Murdoch's reconstruction of the *cybus elementorum*

construction seems very reasonable: presenting a clearly three-dimensional object with easily recognizable analogies to the properties and the four elements. Moreover, as we shall see below, there may be a connection between the cube presented by Isidore and the *cubus perfectus* referenced in the last sentence of the Old Norse *Algorismus*.

3.6.2 Bede

Some ninety years after the composition of Isidore's *De natura rerum*, the English monk Bede would use that treatise as one of the primary models for his works, *De natura rerum* and *De temporibus*.[84] Born in Northumbria in the year 672 or 673 CE, Bede entered the nearby monastery at the age of seven and would spend most of the rest of his life there. He produced a remarkable number of works, ranging from biblical commentaries to histories to educational treatises. Of these, he is most

84 George Hardin Brown, *A Companion to Bede* (Woodbridge, England: The Boydell Press, 2009), 13, 27; Bede, *Bede On the Nature of Things and On Times*, ed. Calvin B. Kendall and Faith Walllis, trans. Calvin B. Kendall and Faith Wallis (Liverpool: Liverpool University Press, 2010), 7–20.

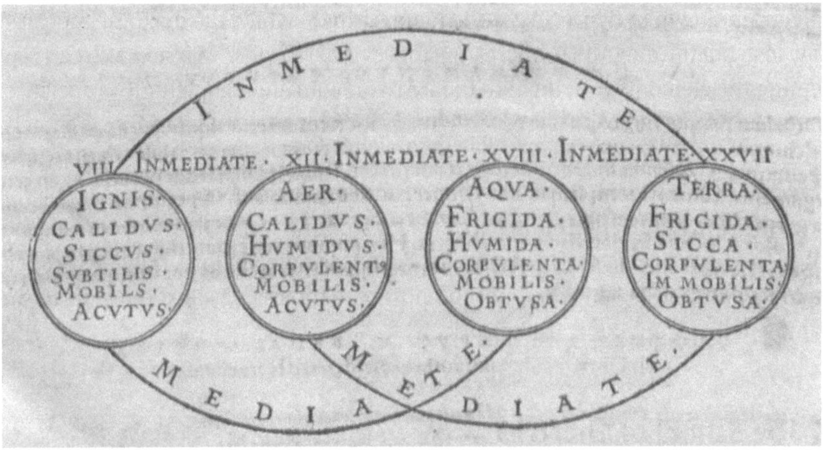

Fig. 3.12: Diagram of the elements in an edition of *De natura rerum*

remembered today for his *Historia ecclesiastica gentis Anglorum*, a history of the Christian church in England, written in 731 CE.[85]

Bede's discussion of the elements in the *De natura rerum* is concise, as is the text as a whole. Moreover, unlike Isidore, his treatment of the nature of the elements refers only to the Aristotelian contraries of dry/wet and cold/hot:

> Quae tamen quadam naturae propinquitate sibimet ita commiscuntur, ut terra quidem arida et frigida frigidae aquae, aqua uero frigida et humida humido aeri, porro aer humidus et calidus calido igni, ignis quoque calidus et aridus terrae societur aridae. Vnde et ignem in terris et in aere nubila terrenaque corpora uidemus.[86]
>
> [These elements nevertheless are so mingled with each other by a certain affinity of nature that dry and cold earth is associated with cold water; cold and moist water is associated with moist air; moist and hot air is associated with hot fire; and hot and dry fire is associated with earth. Hence we not only see fire in the earth but also clouds and earthy bodies in the air.[87]]

However, later scribes and commentators would make the connection to the neoplatonic characterization of the elements. For example, in his printed edition of the collected works of Bede in 1563, Johann Herwagen, known as Hervagius, includes

85 Brown, *A Companion to Bede*, 1–16.

86 Bede, "De natura rerum," in *Beda Venerabilis: Opera didascalica*, ed. Charles W. Jones, vol. I (Brepols, 1975), IIII, 195–96.

87 Bede, *Bede on the Nature of Things*, 4, 76.

the diagram shown in figure 3.12.[88] The manuscript in which the diagram appeared is now lost, but the diagram is interesting for several reasons.[89] First, the circles representing the elements contain both their Aristotelian and neoplatonic properities; second, the properties of air agree with the properties as given by Isidore—namely, air is listed as *corpulentus* (corporeal) and *acutus* (sharp), rather than *subtilis* (fine) and *obtusus* (blunt); and, third, the geometric sequence has been reversed, with 8 representing fire, 12 representing air, 18 representing water, and 27 representing earth, as we saw previously for figure 3.8. Again, as noted before, only the proportionality between the numbers, not the numbers themselves, are relevant to the meaning of the neoplatonic framework.

3.7 The School of Chartres

Over the last fifty years or so there has been an ongoing debate and reassessment of the stature and influence of the cathedral school of Chartres. Whatever the exact details may be, it is evident that the school was central to a revival of neoplatonism in the 12th century.[90] Some of this renewed interest is reflected in commentaries on the *Timaeus*, but can also be seen in other works, including commentaries on the *De institutione arithmetica* and *De consolatione philosophiae* of Boethius. We will look at three prominent members of the school: Bernard of Chartres, Thierry of Chartres, and William of Conches.

3.7.1 Bernard of Chartres

No matter what the exact nature and prominence of the cathedral school of Chartres was in the 12th century, it is clear that Bernard of Chartres was teaching there in the first quarter of that century and that William of Conches was among his disciples.[91] Moroever, both wrote glosses on the *Timaeus*, working from very similar, but not

88 Bede, "De natura rerum," in *Operum Venerabilis Bedae presbyteri*, ed. Johann Herwagen, vol. II (Basel: Hervagius, 1563), IIII, 5.

89 For a discussion of the manusripts used by Herwagen, see: Michael Lapidge, "Byrhtferth of Ramsey and the 'Glossae Bridferti in Bedam'," *The Journal of Medieval Latin* 17 (2007): 384–400.

90 For an overview of the debate, see: Édouard Jeauneau, *Rethinking the School of Chartres*, trans. Claude Paul Desmarais (Toronto: University of Toronto Press, 2009).

91 Paul Edward Dutton, "The Uncovering of the Glosae Svper Platonem of Bernard of Chartres," *Mediaeval Studies* 46 (1984): 192–93.

identical, copies of the translation and commmentary of Calcidius.[92] Most probably, William knew Bernard's commentary, and used it in composing his own.[93]

Very little is known about the life of Bernard. Indeed, the details of his life are so obscure that, for periods in the 18th and 19th centuries, he was confused with two other Bernards: Bernard Silvester and Bernard of Moëlan.[94] We know that he was at Chartres no later than 1114 and he was master of the cathedral school by 1119. He was later chancellor, a post he left in 1126 due either to his retirement or death.[95] The record of his life may be modest, but his reputation as a teacher was held in the highest regard. John of Salisbury, himself a student of William of Conches,[96] wrote that Bernard was "exundantissimus modernis temporibus fons litterarum in Gallia"[97] ["the most abundant fount of literary knowledge in France in modern times"[98]], as well as "perfectissimus inter Platonicos saeculi nostri"[99] ["the most accomplished Platonist of our time"[100]].

The *Glosae super Platonem* of Bernard is thought to date from 1100 to 1115.[101] Only in recent years has a convincing argument for the attribution of the text to Bernard been made. Previously, some considered that it could possible be an early rendition of the gloss of William of Conches.[102] In commenting on *Timaeus* 31 and 32, Bernard, following Calcidius, first lists the contrary properties of earth and fire:

> Ignem et terram uocat prima fundamenta, quia in omni sua proprietate sunt contraria, ita dico si secundum cubos iungantur. Nam terra est corpulenta, obtusa, immobilis; ignis subtilis, acutus, mobilis; et ex his proprietatibus conficiuntur media.[103]

---

92 Bernard of Chartres, *The Glosae Super Platonem of Bernard of Chartres*, ed. Paul Edward Dutton (Toronto: Pontifical Institute of Mediaeval Studies, 1991), 132–33.

93 Dutton, "The Uncovering of the Glosae Svper Platonem," 202–6.

94 Bernard of Chartres, *The Glosae Super Platonem*, 22; Alexandre Clerval, *Les écoles de Chartres au Moyen-Age* (Chartres: Librairie R. Selleret, 1895), 158.

95 Dutton, "The Uncovering of the Glosae Svper Platonem," 193; Bernard of Chartres, *The Glosae Super Platonem*, 32.

96 John of Salisbury, *Metalogicon*, ed. and trans. J. B. Hall (Turnhout: Brepols, 2013), 21–22.

97 Ioannis Saresberiensis, *Metalogicon*, ed. J. B. Hall (Turnhout: Brepols, 1991), I.24, 52.

98 John of Salisbury, *Metalogicon*, I.24, 175.

99 Ioannis Saresberiensis, *Metalogicon*, IV.35, 173.

100 John of Salisbury, *Metalogicon*, IV.35, 330.

101 Bernard of Chartres, *The Glosae Super Platonem*, 15.

102 Dutton, "The Uncovering of the Glosae Svper Platonem"; Bernard of Chartres, *The Glosae Super Platonem*, 8–21.

103 Bernard of Chartres, *The Glosae Super Platonem*, 167.

> [He calls fire and earth the principal foundations, because they are contrary in all of their properties, so I say they are joined according to cubes. For earth is corporeal, compact, immobile; fire is fine, sharp, mobile; and the mediate elements are constructed from these properties.]

He follows this with the standard argument that plane numbers require only one intermediate to join them in geometric progression, but solid numbers require two intermediaries. But, unlike Calcidius, he provides the numerical example of 4, 6, 9 for plane numbers, followed by a detailed explanation of the the progression between the cubes 8 and 27 and the analogous progression through the elements:

> Cubici numeri sunt qui habent tres dimensiones, ut bis bini bis et ter terni ter. Qui iunguntur duobus mediis sumptis ex duobus extimis, ita ut unum medium duo brachia extendat ad unum extimum, tercium ad alterum, et ponatur iuxta illud cuius habet duas partes, ut bis bini ter et alterum similiter, et dicatur, ter terni bis.
> Verbi gratia. Sint duo cubi, VIII et XXVII, differentes per omne latus. Hos iungunt XII (qui sumit bis bini, id est longum et latum ab VIII et ter, id est altum, a XXVII) et XVIII (qui sumit ter terni a proximo, bis a remoto), et iunguntur omnes in sesquialtera habitudine. Sicut ergo inuenitur in cubicis numeris, ita deus, communicando qualitates terrae et ignis aeri et aquae mediis, fecit solidam conexionem. Aqua enim a terra, cui proxima est, duas recipit, scilicet quod corpulenta, quod obtusa; ab igne, quod mobilis. Aer vero ab igne duas recipit, scilicet mobile et subtile; terciam a terra, quod obtusus, unde aues pendent in aere.[104]
>
> [Cubic numbers are those which have three dimensions, such as twice two twice and thrice three thrice. These are joined by two means taken from the two extremes, so that one mean includes two powers from one extreme, and a third from the other, and it is placed next to that of which it has two parts, so twice two thrice and the other similarly, that is to say, thrice three twice.
> For example. Let two cubes, 8 and 27, be different on every side. 12 (which takes twice two, that is, length and width from 8 and a third, that is height, from 27) and 18 (which takes thrice three from the nearest, and two from the furthest) joins these, and all are joined in a sesquialtera relationship. Therefore, as is found in cubic numbers, so God, communicating the qualities of earth and fire with intermediates air and water, made an unbroken connection. For water receives from earth, which is nearest, two, namely, corporeality and compactness; from fire, mobility. Moeover, air recieves from fire two, namely, mobility and fineness; a third from earth, which is compactness, whence the birds hang in the air.]

To this, Bernard adds a comment that Macrobius had previously described another way to connect the elements using only two qualities, but he does not go into any details, instead concluding that the teaching of Plato suffices.[105] Hence it appears that Bernard understands that the characterization of the elements using the dry/wet and cold/hot contraries is not of platonic origin.

---

104 Bernard of Chartres, *The Glosae Super Platonem*, 167–68.
105 Bernard of Chartres, *The Glosae Super Platonem*, 168.

3.7.2 Thierry of Chartres

Few details are known about the life of Thierry of Chartres. He was born in Brittany sometime near the end of the 11th century.[106] He was chancellor of the cathedral at Chartres from approximately 1142 to 1150, succeeding Gilbert of Poitiers who had himself succeeded Bernard in 1126.[107] His disciple, Clarembald of Arras, praised him as "totius Europae philosophorum praecipuus"[108] [the foremost of the philosophers of all Europe].

In recent years, a commentary on the *De institutione arithmetica* of Boethius, probably written in the fifth decade of the 12th century, has been added to the list of works credited to Thierry.[109] In commenting on plane and solid numbers, Thierry expands upon the passage in the *De institutione arithmetica* to make the connection to the elements intermediate to earth and fire:

> Dicit enim Plato quod duo plana una medietate tantum iungantur, et est intelligendum de quadratis superficiebus, quia in caeteris falleret. Dicit etiam quod duo solida duabus medietatibus semper iunguntur, et est intelligendum de cubicis. Medietates vero geometricas debemus intelligere. Unde Plato probat quod ignis et terra duobus mediis necessario sint coniuncta. Facit autem quaestionem Calcidius: quomodo, cum terra sit cubica, ignis vero pyramoides, quomodo inquam duobus mediis coniungantur necessario, cum hoc de cubicis tantum dicatur? Dicit autem postea quod ignis, quamvis sit pyramicus, in mundi tamen compositione rotundus est, et propter rotunditatis perfectionem dicitur cubicus. Unde bene dictum est quod ipse et terra duobus mediis iunguntur, quod in numeris hic ostenditur. Bis bini enim primus quadratus actu est, ter terni secundus, sed bis bini IIII fiunt, ter terni VIIII. Sumatur unum latus quaternarii, id est bis, et aliud latus novenarii, id est ter, et dicatur bis tres, fient VI, qui ad utrumque habet se geometrica medietate. Item bis bini bis primus est cubicus, id est VIII, ter terni ter secundus, id est XXVII. Sumantur ergo duo latera VIII, id est bis bini, et unum latus XXVII, <id est> ter, et dicatur bis bini ter, erunt XII. Sumutur ergo duo latera etiam XXVII, id est ter terni, et unum latus de VIII, id est bis, et dicatur ter terni bis, fient XVIII. Et erit geometrica medietas: XXVII, XVIII, XII, VIII.[110]

> [For Plato says that two plane numbers are joined by one intermediate, and that is to be understood about square surfaces, for that may be mistaken in others. Indeed, he says that two solid numbers are always joined by two intermediates, and that is to be understood of cubes. Moreover, we must understand these to be geometric intermediates. Whence Plato shows that fire and earth are necessarily connected by two means. However, Calcidius asks the question:

---

106 Thierry of Chartres, *The Commentary on the De arithmetica of Boethius*, ed. Irene Caiazzo (Toronto: Pontifical Institute of Mediaeval Studies, 2015), 6.

107 É. Jeauneau, *Rethinking the School of Chartres*, 57, 65.

108 Clarembald of Arras, "Epistola ad Dominam," in *Life and Works of Clarembald of Arras*, ed. Nikolaus M. Häring (Toronto: Pontifical Institute of Mediaeval Studies, 1965), 226.

109 Thierry of Chartres, *The Commentary on the De arithmetica of Boethius*, 22.

110 Thierry of Chartres, *The Commentary on the De arithmetica of Boethius*, II.XLVI, 203–4.

> how, since earth is a cube, and fire a pyramid, how is it said they are necessarily connected by two means, as this is said only about cubes? However, afterwards he says that fire, although it is a pyramid, nevertheless in the composition of the world it is round, and because it is said the perfection of roundness is cubical. Whence that itself is well said and earth is joined by two means, which is shown is accordance with numbers. For twice two is actually the first square, thrice three the second, but twice two makes 4 and thrice three makes 9. Let us take one side from four, that is two, and another side from nine, that is three, and that is called twice three, which makes 6, which is related to the sides of both as the geometric mean. Similarly, twice two twice is the first cube, that is 8, thrice three thrice is the second, that is 27. Then let us take two sides from 8, that is twice two, and one side from 27, <that is> three, and that is called twice two three, which is 12. Then let us take two sides from 27, that is thrice three, and one side from 8, that is two, and that is called thrice three twice, which is 18. And that will be the geometric mean: 27, 18, 12, 8.]

Several things are noteworthy in this passage. First, Tierry shows a concern with the differing descriptions of the elements given by Plato, here using cubes and elsewhere using the Platonic solids. This concerned Calcidius as well, and we will discuss this question further in chapter 9. Second, although Thierry connects the geometric sequence to the elements, and directly references Calcidius, nevertheless he does not mention the description of the elements in terms of the three sets of neoplatonic contraries. Earlier in the commentary he does connect the elements with the Aristotelian contraries: “Ignis terrae siccitate coniungitur, terra aquae frigiditate, aqua aeri humiditate”[111] [Fire is joined to earth by dryness, earth to water by coldness, water to air by wetness]. Similary, Thierry's student Clarembald of Arras, in his commentary on the book of Genesis, describes the elements in terms of the Aristotelian contraries: “Mundus iste ex contrariis elementis coniunctus est: calidis, frigidis, humidis, siccis”[112] [This world is joined together from contraries: hot, cold, wet, dry]. It is worth noting, however, that elsewhere Clarembald shows that he is aware of the description of the elements in terms of the neoplatonic contraries when he writes that the dissimilitude of fire and earth is shown through their

> per omnia proprietates ostendens contrarias ut in igne subtilitatem, acuitatem, mobilitatem; in terra vero obtusitatem, corpulentiam, immobilitatem, non nisi mediorum interiectione connectuntur.[113]

---

111 Thierry of Chartres, *The Commentary on the De arithmetica of Boethius*, I.II, 120.
112 Clarembald of Arras, “Tractatulus super librum Genesis,” in *Life and Works of Clarembald of Arras*, ed. Nikolaus M. Häring (Toronto: Pontifical Institute of Mediaeval Studies, 1965), 9, 229–30.
113 Clarembald of Arras, “Expositio super librum Boetii De Hebdomadibus,” in *Life and Works of Clarembald of Arras*, ed. Nikolaus M. Häring (Toronto: Pontifical Institute of Mediaeval Studies, 1965), 24, 204.

> [displaying contraries through all properties, as fineness, sharpness, mobility in fire; however compactness, corporeality, immobility in earth, are joined only by the insertion of means.]

So it is likely that Thierry was aware of both descriptions of the elements, but may not have distinguished one set being of neoplatonic origin and the other coming from Aristotle.

3.7.3 William of Conches

William of Conches was born in Normandy around the year 1090. He was a student of Bernard of Chartres and, in turn, a teacher of John of Salisbury, most likely at Chartres as well. Although William is best known for his writings in philosophy and natural science, John refers to him as "grammaticus post Bernardum Carnotensem opulentissimus"[114] [the best-endowed grammarian after Bernard of Chartres[115]]. His teaching and writing career began in the third decade of the 12th century, and he most likely remained at Chartres until he returned to Normandy in the 1140s and lived there until at least 1154.[116]

Early in his career, William wrote a commentary, the *Glosae super Boetium*, on the *De consolatione philosophiae* of Boethius. We will discuss William's gloss on Boethius's description of the elements when we look at the *De consolatione philosphiae* later in this chapter. For now, we consider two works of William which relate directly to the presentation of the neoplatonic theory of the elements: the *Glosae super Platonem* (his commentary on the *Timaeus* as witnessed in the translation of Calcidius), and the *Dragmaticon philosophiae*. Both of these works date from the latter years of his career, with the former written sometime around 1140 and the latter composed later in the same decade.[117]

In his gloss of *Timaeus* 31 and 32, after first noting that between plane numbers there is a single geometric mean, using the standard example of 6 being the geometric mean of 4 and 9, William describes the two means between solid numbers:

---

114 Ioannis Saresberiensis, *Metalogicon*, I.5, 20.

115 John of Salisbury, *Metalogicon*, I.5, 135.

116 For more on the life of William of Conches, see: Guillelmi de Conches, *Dragmaticon philosophiae*, ed. Italo Ronca (Turnhout: Brepols, 1997), XVII–XIX; É. Jeauneau, *Rethinking the School of Chartres*, 43–55.

117 For the dating of the works of William of Conches, see: Guillelmi de Conches, *Guillelmi de Conchis: Glosae super Platonem*, ed. Édouard A. Jeauneau (Turnhout: Brepols, 2006), XIX–XLI; Guillelmi de Conches, *Dragmaticon philosophiae*, XIX–XXII.

> Solidus uero est numerus habens tres aequales uoces in sui multplicatione ut bis bini bis qui sunt octo, et ter terni ter id est uiginti septem, dictus solidus propter similitudinem quam habet cum solidis. Habet enim tres aequales uoces ut solida tres aequales dimensiones. Tribus etenim modis numeri referuntur ad res: quantitate, similitudine, dispositione. Inter hos numeros nullus potest inueniri qui aequaliter ad duo extrema se habeat. Sunt igitur duo in medio, id est duodecim et duodeuiginti, qui omnes sunt in sesquialtera proportione.[118]
>
> [But a solid number is a number having three equal terms in its multiplication, such as twice two twice, which is eight, and thrice three thrice, that is, twenty-seven. It is called solid because of the similarity which it has with solid objects. For it has three equal terms as a solid object has three equal dimensions. Indeed, numbers are referred to things in three ways: magnitude, similarity, order. Between these numbers, none can be found which has equal measure to the two extremes themselves. Hence there are two intermediates, that is, twelve and eighteen, which are all in sesquialtera proportion.]

It is of note that William adds an explanation as to why some numbers are called solid numbers: a solid number does not refer directly to a solid cube of geometry, but rather refers through similarity to the equality of the number of factors of a solid number to the number of dimensions of a solid geometric cube. Hence the discussion of the elements is not about their geometric shape, but, via an analogy, concerns the factors which contribute to their composition. This addresses a concern, mentioned above, raised by Thierry and to which we will return in chapter 9.

Using this reference by similarity, William now argues that if two things are contrary in two qualities, then only one intermediate is needed to join them, as water lies between earth and air. But if they differ in three qualities, then two intermediaries are necessary:

> Sed si contrarietas est in tribus, duo media sunt necessaria, ut supra docuimus. Et notandum quod coniunctio contrariorium in duabus qualitatibus plana dicitur sinzugia propter similitudinem cum plano, id est superficie que duas habet dimensiones, siue illa sint solida corpora siue non. Coniuncto vero contrariorum in tribus qualitatibus solida dicitur sinzugia quia solida tres habent dimensiones, ut coniuncto terre et ignis dicitur solida, terrae et aeris plana, etsi illa nichilominus sint solida.[119]
>
> [But if the contrariety is in three ways, two means are necessary, as we showed above. And it should be noted that the union of things contrary in two qualities is called a plane syzygy because of the similarity with a plane, that is a surface which has two dimensions, whether or not they are solid bodies. And the union of things contrary in three qualities is called a solid syzygy because solids have three dimensions, as the union of earth and fire is called solid, earth and air plane, even if they are nevertheless solids.]

---

118 Guillelmi de Conches, *Guillelmi de Conchis: Glosae super Platonem*, LXIII, 110–11.

119 Guillelmi de Conches, *Guillelmi de Conchis: Glosae super Platonem*, LXIII, 111.

William now lists the three contrary qualities of earth and fire, using the same language as Bernard:

> Ergo inter terram et ignem duo media fuerunt necessaria: eorum enim contrarietas in tribus est qualitatibus. Est enim terra obtusa, corpulenta, immobilis; ignis acutus, subtilis, mobilis.[120]
>
> [Therefore between earth and fire two means were necessary: for they have contraries in three qualities. For earth is compact, corporeal, immobile; fire is sharp, fine, and mobile.]

And so William now states his conclusion about the creation of the world:

> ICCIRCO, scilicet quia nulla duo sine medio possunt conuenire, nec solida sine duobus, mundoque opus erat soliditate, OPIFIX MVNDI id est Creator INSERVIT AQUAM ET AERA INTER IGNEM ET TERRRAM, loco et qualitatibus.[121]
>
> [Therefore: namely because no two things are able to be brought together without a mean, nor solids without two means, and there was a need for the world for solidity, the ARTISAN OF THE WORLD, that is, the Creator, INSERTED WATER AND AIR BETWEEN FIRE AND EARTH, in place and in qualities.]

He then expands on the proportionalities needed to progress from earth to fire: water shares two qualities with both earth and air and, in turn, air shares two qualities with both water and fire.

Although William does not refer to any diagrams representing the relationship between the elements, we find diagrams in the marginalia of some later manuscripts. For example, the 13th-century manuscript Conventi Soppressi E.VII 1398 in the Biblioteca Nazionale di Firenze, folio 1v, has a very simple diagram, similar to the one we saw earlier in figure 3.4, as shown in figure 3.13.[122]

William provides a full neoplatonic explication of the necessity of four elments in the *Dragmaticon philosophiae*, written sometime after his gloss on the *Timaeus*. The philosopher in this dialogue begins the story of the creation of the elements by quoting Plato's contention that fire is necessary to make things visible, and earth to make them tangible. After explaining why there must exist intermediaries between fire and earth, and why neither water nor air alone would suffice to mediate between the two, the philosopher explains how two intermediate elements are needed:

---

120 Guillelmi de Conches, *Guillelmi de Conchis: Glosae super Platonem*, LXIII, 111.

121 Guillelmi de Conches, *Guillelmi de Conchis: Glosae super Platonem*, LXIV, 112.

122 Conventi Soppressi E.VIII 1398 (Florence: Biblioteca Nazionale Centrale di Firenze); See: Toni Schmid, ed., "Ein Timaioskommentar in Sigtuna," *Classica et Mediaevalia: Revue Danoise de Philologie et d'Histoire* X, no. 2 (1949): 247.

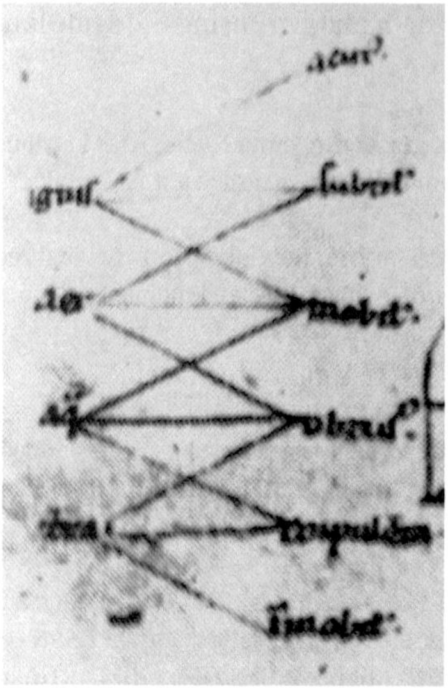

Fig. 3.13: Diagram of the elements in Biblioteca Nazionale E.8 1398, f. 1v

> Duo in medio sunt creata, quorum utrumque a proximo recipit duas de praedictis qualitatibus, a remoto unam: est enim aqua corpulenta obtusa mobilis, aer uero subtilis obtusus mobilis. In hac igitur solida sinzugia corpora praedicta concordant: quae enim uicina sunt duas habent communes qualitates, in tercia differunt.[123]
>
> [Two bodies have been created in the middle {that is, between the two extremes}, each of which recieves, out of the six aforementioned qualities, two from the nearest and one from the farthest body: for water is dense, obtuse, mobile; air thin, obtuse mobile. Thus the aforementioned bodies are in harmony in this solid {that is, cubic threefold} syzygy: for those which are close to each other have two common qualities, but differ in the third.[124]]

When pressed for more clarity, the philosopher, somewhat reluctantly, explains that "[p]ythagoricorum fuit haec consuetudo, quod erat rerum attribuere numeris, et quod erat numerorum attribuere rebus" ["[i]t was the custom of the Pythagoreans to attribute the properties of things to numbers and the properties of numbers to things"], and that "Plato uero, qui pythagoricus fuit, in illo capitulo quod est in numeris mundo attribuit; cuius haec est ratio: numquam solidus numerus potest uno medio uinciri, planus uero semper uno medio uincitur"[125] ["Plato, who was a

123 Guillelmi de Conches, *Dragmaticon philosophiae*, II.5.1, 45.
124 Willliam of Conches, *A Dialogue on Natural Philosophy*, trans. Italo Ronca and Matthew Curr (Notre Dame, IN: University of Notre Dame Press, 1997), II.5.1, 29–30.
125 Guillelmi de Conches, *Dragmaticon philosophiae*, II.5.4, 47.

Pythagorean, attributes the properties of numbers to the world, and his argument is as follows: a solid number can never be bound by a single mean, whereas a plane number is always bound be a single mean”[126]]. In more detail:

> Solidus est numerus qui habet tres aequales uoces in sui multiplicatione, ut bis bini bis, qui sunt VIII, et ter terni ter, qui sunt XXVII. Planus uero est numerus qui duas tantum habet aequales uoces in sui multplicatione, ut bis bini, qui sunt IV, et ter terni, qui sunt IX. Vincire uero numeros est numerum infra illos inuenire qui eadem proportione habeat se ad minorem, qua maior ad ipsum. Hoc in planis numeris semper inuenitur, quemadmodum inter quaternarium et nouenarium senarius. Sed inter duos solidos nullus talis esse potest, sed duo inueniuntur semper, ut inter VIII et XXVII sunt XII et XVIII, sic: VIII-XII-XVIII-XXVII. Omnes isti sunt in sesqu<i>altera proportione.
> Quia igitur numeri qui habent soliditatem uno medio uinciri non possunt, dixit Plato, attribuendo rebus quod est numerorum, solida sine duobus mediis uinciri non posse.[127]

> [A solid number is one that has three equal figures in the multiplication of itself, as [2^3 =] $2 \times 2 \times 2 = 8$ and [3^3 =] $3 \times 3 \times 3 = 27$. A plane number is one that has only two equal figures in the multiplcation of itself, as [2^2 =] $2 \times 2 = 4$ and [3^2 =] $3 \times 3 = 9$. To bind two numbers means to find a number between them that is related to the smaller in the same proportion as is the greater to itself. Such a number is always found in the case of plane numbers, as the number 6 [is the geometric mean] between the numbers 4 and 9. Between two solid numbers, however, there can be no such single mean, but always two are found, as 12 and 18 between 8 and 27, in this progression: 8 – 12 – 18 – 27. All these numbers are successively proportionate by the ratio of $1\frac{1}{2}$.
> Since, then, two numbers with solid structure canot be conjoined by a single mean, Plato, by attributing to things what is proper to numbers, maintained that solid things cannot be conjoined without two means.[128]]

Once again, we see William being careful to note that Plato, as a pythagorean, is directly identifying properties of numbers with properties of physical things, whereas the reference of numbers to things should be through similarity, not identity.

In the course of this discussion, William also mentions the harmony between the elements in terms of other qualities, including the dry/wet and cold/hot contraries we have seen before:

> Similiter in plana sinzugia concordant: est enim terra frigida et sicca, aqua frigida et humida, aer calidus et humidus, ignis calidus et siccus.[129]

126 Conches, *A Dialogue on Natural Philosophy*, II.5.4, 31.
127 Guillelmi de Conches, *Dragmaticon philosophiae*, II.5.5–6, 47–48.
128 Conches, *A Dialogue on Natural Philosophy*, II.5.5–6, 31–32.
129 Guillelmi de Conches, *Dragmaticon philosophiae*, II.5.1, 45.

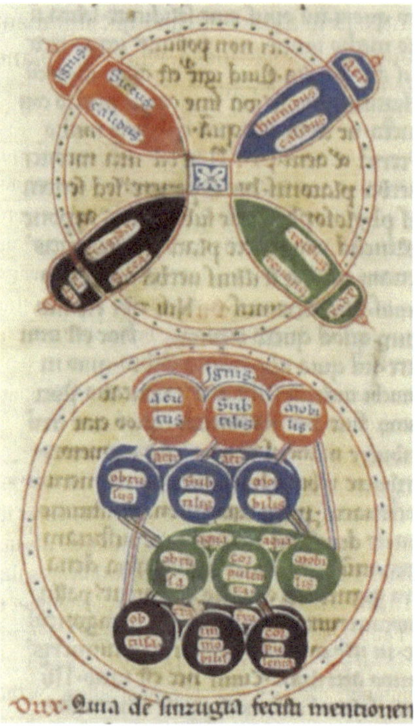

Fig. 3.14: Diagram of the elements in the *Dragmaticon* in BNF MS Latin 6415, f. 6r

> [Similary, they are in harmony in the plane {that is, square, twofold} syzygy for earth is cold and dry, water cold and moist, air hot and moist, and fire hot and dry.[130]]

Noting that images are useful for remembering such relationships, he provides a diagram. Figure 3.14 is an example of such a diagram from the 14th-century manuscript Latin 6415 in the Bibliothèque nationale de France.[131] More directly related to this discussion are the illustrations included at the end of the dialogue in the 13th-century manuscript Bodleian Library MS Digby 107, folio 51r. One of these, shown in figure 3.15, is a more elaborate version of the diagram representing the

---

130 Conches, *A Dialogue on Natural Philosophy*, II.5.1, 30.

131 "*Magistri Guillelmi de Conchis* opus inscriptum: Dragmaticon philosophiae," in *BNF Latin 6415* (Bibliothèque nationale de France), f. 6r, accessed March 7, 2026, https://archivesetmanuscrits.bnf.fr/ark:/12148/cc655009?collect.

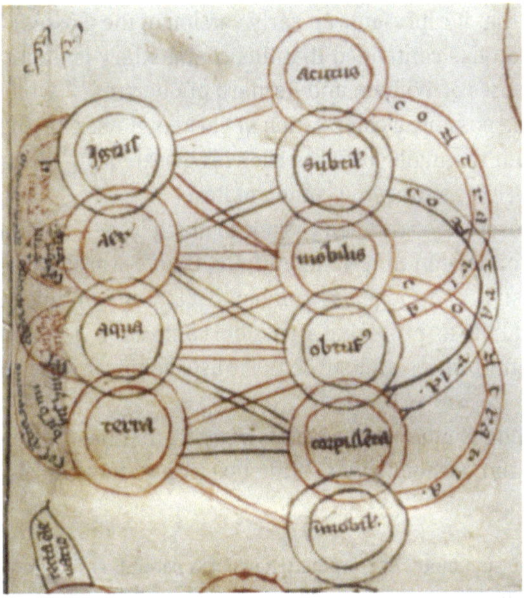

Fig. 3.15: Diagram of the elements in the *Dragmaticon* in Digby 107, f. 51r

relationships between the elements that we saw previously in figure 3.8.[132] This is the same diagram (shown in figure 3.18 below) that appears in AM 736 III, the manuscript which contains a fragment of the Old Norse *Algorismus*. Although that fragment of the *Algorismus* does not contain the discussion of the elements, the inclusion of this diagram may indicate a connection between the works of William of Conches and the concluding section of the Old Norse *Algorismus*.

3.7.4 The Uppsala Manuscript

The manuscript Universitetsbiblioteket MS. C 620 in Uppsala contains a commentary of uncertain authorship on the *Timaeus*.[133] This 13th-century manuscript is known to have once been part of the library of the Dominican friary in Sigtuna, Swe-

132 "Summa super naturalibus quaestionibus," in Bodleian Library MS. Digby 107 (Bodleian Library, University of Oxford), accessed December 10, 2025, https://digital.bodleian.ox.ac.uk/objects/71deba7a-66d2-4d92-8647-25421da56794/surfaces/e54e853c-b91c-48f8-9f5a-d3e9610d7a84/.
133 "Commentarius in Timaeum Platonis," in Universitetsbiblioteket MS. C 620 (Uppsala: Uppsala Universitetsbibliotek); Schmid, "Ein Timaioskommentar in Sigtuna."

den.[134] While some have argued that it represents an early version of the glosses of William of Conches,[135] P. E. Dutton has contended that this commentary is the result of a scribe combining the glosses of William and Bernard of Chartres,[136] while Édouard Jeauneau thought it to be possibly a compilation of various glosses, saying that "[l]es plagiats, on le sait, sont monnaie courante chez les glosateurs"[137] [plagarism, we know, is common currency among the commentators].

The importance of this manuscript for this story is not so much its description of the elements, as its possible connection to Iceland. Although there were no Dominican friaries in Iceland, Christian Etheridge has pointed to a strong Dominican influence on Icelandic intellectual life as providing a possible connection between manuscripts found in Dominican libraries and subsequent astronomical texts in Iceland.[138]

After discussing how other philosophers have erred in saying that the four elements were created indistinct and without order,[139] the text argues that the elements are, by nature, distinct;

> Elementorum naturaliter contrarii sunt motus et inordinati. Nam terra naturaliter fertur deorsum ignis sursum aqua in longum. aer in latum.[140]

> [The motions of the elements are inherently contrary and inordinate. For earth is naturally carried downward, fire upward, water in length, air in breadth.]

As in the *Timaeus* and subsequent commentaries, earth and fire are first among the elements in that they are needed to make things tangible and visible. And, as earth and fire differ in all of their properties, other elements are needed to mediate between them:

---

134 Christian Etheridge, "The Context of the Astronomical Treatises in GKS 1812 4to," in *A World in Fragments: Studies on the Encyclopedic Manuscript GKS 1812 4to*, ed. Gunnar Harðarson et al. (Reykjavík: Stofnun Árna Magnússonar í Íslenskum Fræðum, 2021), 250.

135 For examples, see: Schmid, "Ein Timaioskommentar in Sigtuna," 220–25; R. W. Southern, *Platonism, Scholastic Method, and the School of Chartres* (Reading, England: University of Reading, 1979), 16.

136 Dutton, "The Uncovering of the Glosae Svper Platonem," 201–2.

137 Édouard A. Jeauneau, "L'usage de la notion d'*integumentum* a travers les gloses de Guillaume de Conches," *Archives d'Histoire Doctrinale et Littéraire du Moyen Age* 23 (1957): 58.

138 Etheridge, "The Context of the Astronomical Treatises," 256–57.

139 "Commentarius in Timaeum Platonis," f. 84v; Schmid, "Ein Timaioskommentar in Sigtuna," 238.

140 "Commentarius in Timaeum Platonis," f. 85r, ll. 4–5; Schmid, "Ein Timaioskommentar in Sigtuna," 239.

quia voluit racionabilem creaturam creare deus cui gloriam suam manifestaret. que stare non posset sine soliditate. et non viderat aliquid sine splendore. ideo quia nullus splendor sit sine igne. nichil iterum possit solidum esse sine terra. deum ignem et terram creasse. ignem et terram vocat prima fundamenta. que in omni sua proprietate sunt contraria. et ex suis proprietatibus conficiuntur media. licet tamen sciendum est. non ita deum creasse .ꝶ. elementa per successiones ut primitus ignem et terram.[141]

[Because God wished to create a rational creature, through which his glory would be revealed, and to stand immovable is not possible without solidity and nothing is visible without brightness, therefore, because nothing has brightness without fire and again nothing is able to be solid without earth, God created fire and eath. Fire and earth are called the first foundations and are contrary in all of their characteristics. And means are constructed from their characteristics, although it is nevertheless to be understood that God did not in this manner create the 4 elements in succession when first creating fire and earth.]

For the corresponding numerical representation of the geometrical progression, the text provides only the planar example 4, 6, and 9, while noting then that two intermediaries are needed for solids:

Dixit deum creasse ignem et terram ut mundus esset visibilis sed quia illa plurimum erant Inter se discordia. ideo ostendit duo elementa deum creasse intermedia scilicet aerem et aquam et tales utrique contulisse proprietates. ut et se inuicem et secum extrema vincirent. precipue et ut mundus solidus efficeretur. cum nulla soliditas sine duobus mediis constare possit. et in planis vnum efficiat quod in numeris videri aperte potest.[142]

[He said God created fire and earth so the world would be visible but also because they had the most discord between themselves. Then it is explained that God created two intermediate elements, namely, air and water, such that each of them united the characteristics, both in turn to themselves as well as the extremes with themselves, and as especially the world was made solid, since nothing with solidity is able to abide without two means, even as in a plane one mean effects, which can be seen clearly in numbers.]

Although not providing the full neoplatonic account as witnessed in the works of William of Conches, the manuscript does show that these ideas were present in Nordic monasteries by the 13th century, thus providing a possible route to reach monastic schools in Iceland.

141 "Commentarius in Timaeum Platonis," f. 85v, ll. 8–13; Schmid, "Ein Timaioskommentar in Sigtuna," 241.

142 "Commentarius in Timaeum Platonis," f. 85v, ll. 23–27; Schmid, "Ein Timaioskommentar in Sigtuna," 241.

3.8 The Four Elements in the *Algorismus*

3.8.1 John of Sacrobosco and Peter of Dacia

Although we know of no texts treating the Indian calculus which also present the neoplatonic account of the four elements other than the Old Norse *Algorismus*, there are at least two versions which discuss the geometric progression between squares and cubes: the *Algorismus vulgaris* of John of Sacrobosco and the commentary on the *Algorismus vulgaris* written by Peter of Dacia.

From the mid-13th century to the mid-16th century, John's version of the *Algorismus* was the most commonly used text on the subject.[143] Yet so little is known about John himself that he has been described as "a man of mystery but of such great importance that he deserves careful study."[144] He is commonly thought to have been born in England near the beginning of the 13th century, but spent most of his career teaching and writing in Paris. His *Algorismus vulgaris* most likely dates from sometime in the 1220s.[145]

In the preamble to his presentation for extracting square and cube roots, John of Sacrobosco describes three types of numbers: linear (*linearis*), planar (*superficialis*), and solid (*solidus*).[146] The first has only length (*longitudo*), the second length and breadth (*latitudo*), and the last has length, breadth, and thickness (*spissitudo*). He seeks to distinguish among the planar integers those which are squares and among the solid numbers those which are cubes. Here he brings up the arithmetic of Boethius, and the means found between successive squares and cubes:

> Notandum etiam quod inter quoslibet duos quadratos proximos est unicum medium proportionale, quod provenit ex ductu radicis unius quadrati in radicem alterius. Inter duos cubicos quoslibet proximos est duplex medium proportionale, scilicet minus medium et maius. Minus medium provenit ex ductu radicis maioris cubici in quadratum minoris; maius, si ducatur radix minoris cubici in quadratum maioris.[147]

> [Furthermore, it is to be observed that beween any two nearest squares there is one proportional mean, which arises from multiplying the root of one square with the root of the other. Between any two nearest cubes there are two proportional means, namely a smaller mean and

143 Benedict, *A Comparative Study of the Early Treatises*, 12–13.

144 Josiah Cox Russell, *Dictionary of Writers of Thirteenth Century England* (New York: Burt Franklin, 1971), 73.

145 For more on the life and works of John of Sacrobosco, see: Thorndike, *The Sphere of Sacrobosco*, 1–14; Olaf Pedersen, "In Quest of Sacrobosco," *Journal for the History of Astronomy* 16, no. 3 (1985): 175–221.

146 Iohannis de Sacro Bosco, "Algorismus," 192–94.

147 Iohannis de Sacro Bosco, "Algorismus," 194–95.

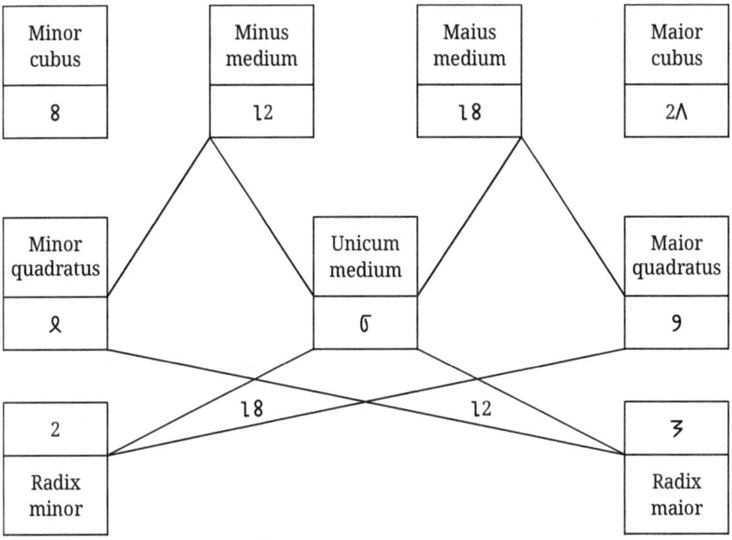

Fig. 3.16: Illustration of the means between squares and cubes

> a larger mean. The smaller mean arises from multiplying the root of the larger cube with the square of the smaller; the larger, if the root of the smaller cube is multiplied with the square of the larger.]

John does not provide a particular example, but the marginalia in at least one early 14th-century manuscript of the *Algorismus vulgaris* illustrates the matter with a diagram using the standard examples of 4, 6, 9 for squares and 8, 12, 18, 27 for cubes, as reproduced in figure 3.16.[148]

Over the centuries, the Danish astronomer and mathematician Peter of Dacia, also known as 'Philomena' and 'Peter Nightingale,' has been confused with various other writers named Peter who lived in Scandinavia in the late 13th and early 14th centuries. Fewer texts are now ascribed to him; still, the ones that remain, covering topics in mathematics, calendar computations, and astronomy and astronomical instruments, were clearly influential for a significant period of time. Although he was a canon in Roskilde, Denmark, he taught and wrote both in Bologna and Paris.[149]

148 "Algorismus sive ars introductorius in numerum," in Kgl. bibl. X 767 (Kungliga biblioteket, Stockholm), 21v; Iohannis de Sacro Bosco, "Algorismus," 169, 195.

149 For more on the life and works of Peter of Dacia, see: F. S. Pedersen, *Petri Philomenae de Dacia et Petri de S. Audomaro*, 26–41; Olaf Pedersen, "Petrus Philomena de Dacia: A Problem of Identity," *Cahiers de l'Institut du Moyen-Âge Gred et Latin* 19 (1976): 1–11; Olaf Pedersen, "The Life and Work of

Peter lectured on the *Algorismus vulgaris* of John of Sacrobosco while in Bologna, and in the summer of either 1291 or 1292, completed his *Expositio super algorismum*, a commentary on John's work. In this commentary, Peter discusses the standard examples for both squares and cubes in some detail. For cubes, he writes:

> Verbi gratia, primus cubicus est 8, secundus proximus est 27, radix primi est 2, radix secundi 3, et ideo sunt proximi cubici, quia inter eorum radices non est medium. Inter istos duos cubicos est duplex medium proportionale, scilicit 18 et 12: sicut enim 27 semel continent 18 et medietatem eorum, sic 18 continent 12 et medietatem eorum, et consimiliter 12 continent 8 et medietatem eorum, ita quod omniquaque est proportio sexquialtera. Et proveniunt haec media sic, sicut dicit auctor, quod *ducendo radicem maioris cubici in quadratum minoris cubici provenit minus medium*; sic scilicet: quia radix maioris cubici erat 3 et quadtratum minoris cubici erat 4, ducendo igitur 3 in 4 exeunt 12, quod est minus medium proportionale; *ducendo etiam radicem minoris cubici*, scilicit 2, *in quadratum maioris cubici*, scilicet in 9, exibit maius medium, scilicet 18.[150]
>
> [For example, the first cube is 8, the next nearest is 27, the root of the first is 2, the root of the second is 3, and so they are nearest cubes, because there is no intermediate between their roots. Between those two cubes there are two proportional means, namely 18 and 12: For just as 27 contains 18 and half of it, so 18 contains 12 and half of it, and similarly 12 contains 8 and half of it, thus what is in every way a sesquialtera proportion. And these means arise so, as the author says, because multiplying the root of the larger cube with the square of the root of the smaller cube creates the mean; namely: because the root of the larger cube was 3, and the square of the root of the smaller cube was 4, therefore multiplying 3 with 4 produces 12, which is the smaller proportional mean; likewise multiplying the root of the smaller cube, namely 2, with the square of the root of the larger cube, namely with 9, will produce 18.]

Moreover, Peter illustrates the relationships with a diagram, although not as elaborate as that seen above in the manuscript of the *Algorismus vulgaris*. This diagram is reproduced in figure 3.17.[151]

Neither John nor Peter mention the relationship connecting the geometric progression between cubes and the elements, yet John's reference to Boethius shows that they are aware of the context in which this question arises.

Peter Nightingale," in *Vistas in Astronomy: New Aspects in the History and Philosophy of Astronomy*, ed. Arthur Beer, vol. 9 (Oxford: Pergamon Press, 1967), 3–10.

150 Petrus de Dacia, "Expositio super Algorismum," 144.

151 Petrus de Dacia, "Expositio super Algorismum," 144.

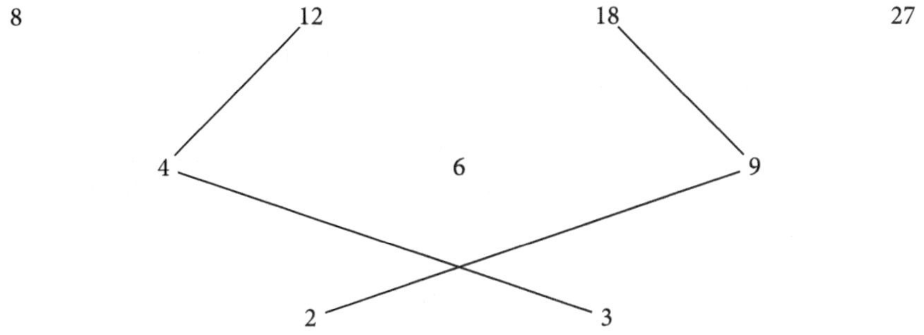

Fig. 3.17: Illustration of the means between squares and cubes

3.8.2 The Old Norse *Algorismus*

Unlike the *Algorismus vulgaris* of John of Sacrobosco and the *Expositio super algorismum* of Peter of Dacia, the Old Norse *Algorismus* gives a full neoplatonic treatment to the geometric progression between cubes. Seemingly out of place to a modern reader, the thread of neoplatonic thought we have sketched above shows how this discussion would not seem so unusual to a 14th-century reader of a treatise on arithmetic. From Nicomachus and Boethius to John of Sacrobosco and Peter of Dacia, it was natural to include explications of the geometric sequence connecting two squares or two cubes in books on arithmetic. Moreover, it was well known that this result was connected to claims made by Plato in the *Timaeus*. What stands out in the Old Norse *Algorismus* is the expansion on this connection to include the full neoplatonic description of the elements. The explication given here is very similar to that of William of Conches in his commentary on the *Timaeus* and in the *Dragmaticon*. But whatever the direct source might have been, the jump from operations of arithmetic to the creation of the elements is not as great a leap as it might at first appear.

The final section begins with a listing of the squares and cubes of the digits from 1 through 9, and then turns to the creation and necessity of the four elements. As with the accounts we have seen above, the discussion hinges on the three-dimensionality of the world:

> Huer ferſkȯ̉yt tala hefuir tuœr mœlínggar þat er breíd ok lengd. En cubicus tala hefuir þrenna mœlíngh. þat er breíd ok lengd ok þycd edir hœd. Oc þui kalla ſpekíngar huern ſynilighan likama með þeſſi tolu ſaman ſettan. at han hefuir. jamnan þeſſa mœlíngh þrenna.[152]

152 GKS 1812 4to, f. 16r, ll. 34–36, f. 16v, ll. 1–3.

[Each square number has two dimensions, that is breadth and length. But a cubic number has three dimensions, that is breadth, length, and thickness or height. And therefore the sages say every visible body is put together with this number for it always has these three dimensions.]

Next we hear again of the need, set out by Plato in the *Timaeus*, that the world must be tangible and visible, and so requires earth and fire, which, being three-dimensional, are identified with the first two cubes:

Með þui at elif ſpeki. oc ein gud uilde heimin ſyinlighan ok likamlighan ſkapa. þa ſetti han fyrſt tuœr ennar yðſtu hofudſkempnnur eld ok jord. þui at ekki ma naturuligha ſynilight vera uttan þœr Þar ſem elddir gerir lioſ ok rǒringh. En jord ſtaðfeſti ok hald. En með þui at þau hafua þrenna víamnna huiligleika ok gangſtadlígha pa. uar naturuligh naudſyn at ſetia nokkot milli þeirra þat er ſamþykttí þeirra vſœtti. Ok ſem fyr er faght at elddir ok jord ok þat alt ſem likamlight er er með þrefaldre tolu er uer kollum cubicum ſaman ſet þa ritum uer þeſſa tua cubus. Ritum uer jordina þeſſa leid. Tuiſuar ſinnum tueir tuiſuar .꜒ꝶ8. En eldin ſua þryſuar þrír þryſ uar .϶9꜒Λ.[153]

[Owing to eternal wisdom, the one God wished to create a visible and bodily world. Accordingly, he first made two of the outermost elements, fire and earth, because nothing can be naturally visible without them, as fire makes light and movement, but the earth is steadfast and unmoving. And since they have three unequal and opposite qualities, it was then a natural necessity to set something between them which reconciled their differences. And since, as said before, fire and earth, and every such thing that is of a bodily nature is composed out of a three-fold number, which we call cubic, then we write for this two cubes. We write the earth this way: twice times two twice, 2, 4, 8. But fire so: thrice three thrice, 3, 9, 27.]

This is followed with the now standard deduction of the necessity of two elements between earth and fire to mediate their differences—namely, water identified with 12 and air with 18—concluding with

Sua ſkipadí gud tuennar hofutſkepnur millí eldz ok íardar. lopt ok vatn. oc hefuir uatn tua huiligleika af íorð oc tuœr tolur. En af elddí eín huiligleik ok eína tolu. En lopt hefuir tua huilighleika af eldí ok tuœr tolur. En eín af íorð ok eina tolu. Oc er elddir þuí lettarí en lopt ſem .꜒Λ. eru meírí en .ɩ8. En lopt þuí lettarí en uatn ſem .ɩ8. eru meírí en .ɩ꜒. vatn þui lettara en jorð ſem .ɩ꜒. eru meirí en .8.[154]

[Thus God arranged two elements between fire and earth: air and water. And water has two qualities from earth and two numbers, but from fire one quality and one number. And air has two qualities from fire and two numbers, but one from earth and one number. And fire is as much lighter than air as 27 is greater than 18. And air is as much lighter than water as 18 is greater than 12. Water is as much lighter than earth as 12 is greater than 8.]

153 GKS 1812 4to, f. 16v, ll. 3–13.
154 GKS 1812 4to, f. 16v, ll. 23–29.

This section, and the *Algorismus* as a whole, now closes with a somewhat enigmatic statement:

> Þetta <er> fullígar at ſkilía j þeirri figuru er her er ſiðar gor oc kollud er cubus PErFectus.[155]
>
> [This is to be more fully understood in that figure which is made here later and is called the *cubus perfectus*.]

It is unclear what figure is referred to here, as there is no corresponding figure in GKS 1812, AM 544, or AM 685 d. However, the statement is similar to what the philosopher says in the *Dragmaticon* immediately before the insertion of a figure illustrating the relationships between the elements:

> Sed quia melius retinentur quae oculis uidentur, utriusque sinzugiae fiat descriptio.[156]
>
> [But since what one sees with one's own eyes is better retained [by the mind], a diagram of each syzygy should be provided.[157]]

Moreover, as noted previously, AM 736 III 4to, although containing only a fragment of the *Algorismus*, a section of the text which does not include the discussion of the elements, does have a diagram illustrating the four elements. This diagram, shown in figure 3.18,[158] is very close to the diagram from a manuscript of the *Dragmaticon* that we saw above in figure 3.15. This may then be the diagram which the *Algorismus* is calling the *cubus perfectus*, providing further evidence of a connection between this final section of the Old Norse *Algorismus* and the *Dragmaticon* of William of Conches. Kristín Bjarnadóttir has previously conjectured that the diagram in AM 736 III may be the *cubus perfectus* missing from the other manuscripts of the *Algorissmus*, although without the evidence provided by the figure in the *Dragmaticon* manuscript.[159]

Yet this figure does not provide any sense of the three-dimensionality that one would expect from the reference to a cube. It may be that a scribe, or whoever translated the text into Old Norse, has mistaken this diagram for other illustrations which are more clearly three-dimensional. An obvious candidate for such a reference to

---

155 GKS 1812 4to, f. 16v, ll. 29–31.

156 Guillelmi de Conches, *Dragmaticon philosophiae*, II.5.1, 45.

157 Conches, *A Dialogue on Natural Philosophy*, II.5.1, 30.

158 "AM 736 III 4to," in *Handrit.is* (Copenhagen: Arnamagnæan Institute), f. 2r, Photograph: Suzanne Reitz, accessed March 2, 2026, https://handrit.is/manuscript/view/da/AM04-0736-III.

159 Kristín Bjarnadóttir and Bjarni V. Halldórsson, "*Algorismus*: Hindu-Arabic Arithmetic in GKS 1812 4to," in *A World in Fragments: Studies on the Encyclopedic Manuscript GKS 1812 4to*, ed. Gunnar Harðarson et al. (Reykjavík: Stofnun Árna Magnússonar í Íslenskum Fræðum, 2021), 183.

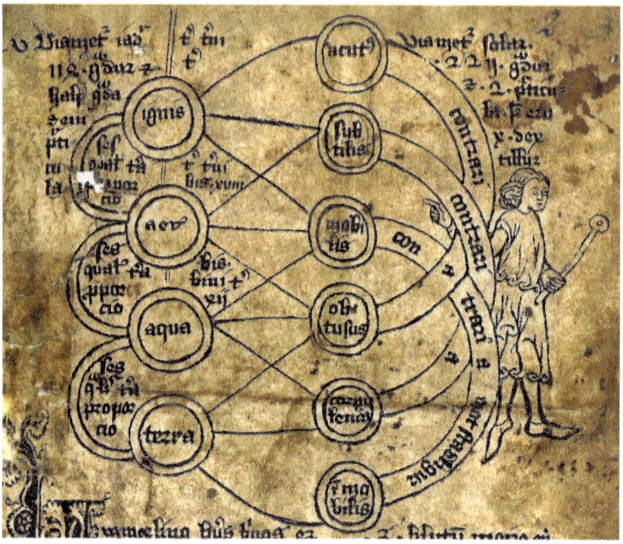

Fig. 3.18: Diagram of the elements in AM 736 III 4to. f. 2r

a *cubus perfectus* might be the *cybus elementorum* of Isidore of Seville (which we saw earlier), a speculation also shared by Otto Bekken.[160] Hence it may be that the diagram in AM 736 III has links to the diagram from the *Dragmaticon* of William of Conches, while the term *cubus perfectus* was originally a reference to something more closely related to the figure seen in the *De natura rerum* of Isidore.

3.9 Other Threads

To this point we have followed a thread of neoplatonic thought, running from late antiquity through the Middle Ages and found in works related to the *Timaeus*, natural science, and arithmetic, which identifies the progression of the elements from earth to water to air to fire with the numeric progression 8, 12, 18, 27 between the first two cubes. But echoes of the *Timaeus* are found in other works as well, especially those dealing with the creation of the world or the calculation of dates.

160 Otto B. Bekken, *On the CUBUS PERFECTUS of the Algorismus in Hauksbók* (Agder, Norway: Agder distriktshøgskole, 1986), 14.

3.9.1 Creation

3.9.1.1 Ovid

In the opening verses of the *Metamorphoses*, composed during the years 2 to 8 CE,[161] Ovid describes how a god brought order out of the chaos of the first creation of the world:

> utque erat et tellus illic et pontus et aer,
> sic erat instabilis tellus, innabilis unda,
> lucis egens aer; nulli sua forma manebat
> obstabatque aliis aliud, quia corpore in uno
> frigida pugnabant calidis, umentia siccis,
> mollia cum duris, sine pondere habentia pondus.
> Hanc deus et melior litem natura diremit.
> nam caelo terras et terris abscidit undas
> et liquidum spisso secreuit ab aere caelum;
> quae postquam euoluit caecoque exemit aceruo,
> dissociata locis concordi pace ligauit.[162]
>
> [And so in that place was the earth and the sea and the air, in such a way that the earth was unsteady, the flood unnavigable, the air wanting of light; nothing with form endured and each opposed the others, because in one body cold fought with hot, wet with dry, soft with hard, heavy with light. A god and better nature broke up this strife. For he tore the earth away from heaven and the flood from the earth and he separated the liquid from the dense, heaven from the air; after that he unfolded the hidden and freed the piles, he bound the parts separated by place with concordant peace.]

Here we see Ovid describing the Aristotelian contrary properties of the elements, cold/hot and dry/wet, to which he adds two more pairs—namely, soft/hard and heavy/light—and how a god brought them from the initial chaos of creation into a harmonious balance.

Later medieval commentators would inevitably interpret this passage in terms of the established neoplatonic scheme. Two manuscripts in the Bayerische Staatsbibliothek provide examples: one in Codex Latinus Monacensis (clm) 4610, and another in clm 14482c. The former is the earliest known commentary on the *Metamorphoses*, written around the beginning of the 12th century, and the second dates from

161 Gian Biagio Conte, *Latin Literature: A History*, trans. Joseph B. Solodow (Baltimore and London: The Johns Hopkins University Press, 1994), 341.
162 Ovid, *Metamorphoses*, ed. R. J. Tarrant (Oxford: Oxford University Press, 2004), I.15–25, 2.

the latter part of the same century.[163] In commenting on the last line quoted above, the anonymous author of the commentary in clm 4610 writes:

> Ignis est acutus, subtilis, mobilis. Aer subtilis, mobilis, obtunsus. Aqua mobilis, obtunsa, corpulenta. Terra obtunsa, corpolenta, immobilis. Aer habet ab igne, quod est subtilis et mobilis, a terra, quod est obtunsus. Aqua habet a terra, quod est obtunsa et corpolenta, ab igne, quod est mobilis.[164]
>
> [Fire is sharp, fine, mobile. Air is fine, mobile, compact. Water is mobile, compact, and corporeal. Earth is compact, corporeal, and immobile. Air has from fire what is fine and mobile, from earth what is compact. Water has from earth what is compact and corporeal, from fire what is mobile.]

The author of the commentary in clm 14482c, also anonymous, expands on this:

> [A]dducentur duo fundamenta penitus oppositas qualitates habentia, uidelicet ignem et terram. Ignis est calidus, acutus, mobilis. Terra uero est frigida, optusa, non mobilis, corpulenta. Hec duo ita sunt repugnantia, scilicet dum per se uidelicet non possunt coherere, posuit quiddam medium, scilicet aquam, que habet duas affinitates cum terra et terciam cum igne. Aqua namque frigida, mobilis, corpulenta. Cum per hoc medium non possit fieri firma concordia terra propter pondus, posuit aliud medium, scilicet aerem unam proprietatem cum terra habentem et duas cum igne. Aer enim calidus, mobilis, corpulens. Et notandum est, quod sicut ignis habet se ad aerem, et ita aer ad aquam, et sicut aer ad aquam, ita aqua ad terram et eodem modo ascendendo, item sicut se ignis habet ad aquam et aer habet se ad terram eodem modo ascendendo.
> Et hoc fecit ad exemplum duorum numerorum cubicorum primum perfectum, scilicet bis bini bis et ter terni ter. Cubiti dicuntur, qui habent longitudinem, latitudinem, spissitudinem, hi duo numeri supradicti, cum essent penitus oppositi. Unus enim constabat ex paribus, scilicet bis bini bis. Alter ex partibus in partibus, scilicet ter terni ter. Non poterat firmiter colligari, nisi interpositis duobus mediis, id est bis bini ter et ter terni bis. Bini ter habet duas affinitates cum bis bini <bi>s, sed terciam cum ter terni ter. Ter terni bis habet duas cum ternis ter et unam cum bis bini bis. Quibus mediis interpositis ita per se colligantur, quod unus se habet ad alium, et ita per certos sexqualtera proportio est.[165]
>
> [Two foundations having completely opposite qualites are brought together, namely, fire and eath. Fire is hot, sharp, mobile. And earth is cold, compact, not mobile, corporeal. So these two are incompatible, since clearly they are not able to cohere naturally by themselves, he {god} made some mean, namely water, which has two affinities with earth and a third with fire.

163 Robin Wahlsten Böckerman, ed., *The Bavarian Commentary and Ovid* (Cambridge: Open Book Publishers, 2020), 170, 133.

164 Robin Wahlsten Böckerman, ed., "Edition and Translation of clm 4610," in *The Bavarian Commentary and Ovid* (Cambridge: Open Book Publishers, 2020), 198.

165 Robin Wahlsten Böckerman, ed., "Bavarian B, clm 14482c," in *The Bavarian Commentary and Ovid* (Cambridge: Open Book Publishers, 2020), 338, 340.

> Indeed water is cold, mobile, corporeal. Since through this mean it is not possible to make a durable union with earth because of weight, he made another mean, namely air having one property with earth and two with fire. For air is hot, mobile, corporeal. And note that just as fire is itself to air, and so air to water, and just as air is to water, so is water to earth and ascending in the same manner, likewise just as fire is itself to water and air is itself to earth ascending in the same manner.
> And he made this according to the example of the first two perfect cubes, namely twice two twice and thrice three thrice. These are called cubic which have length, width, depth, although it is also said these two numbers are completely opposite. For one is constituted from evens, namely twice two twice. The other from odd parts,[166] namely thrice three thrice. He could not steadfastly bind them, except having interposed two means, that is, twice two thrice and thrice three twice. [Twice] two thrice has two affinities with twice two twice, but also a third with thrice three thrice. Thrice three twice has two with thrice three [thrice] and one with twice two twice. So by these means having been interposed they are bound to one another, what one has itself to another, and so there is by certain numbers a sesquilatera proportion.]

The presentation here is somewhat confused; in particular, earth is given four qualities instead of three, the contraries cold/hot are used instead of compact/sharp, and no reference is made to the need for two intermediaries between cubes while one suffices for squares. Still, the author has outlined the basics of the by-now standard neoplatonic description of the binding of four elements.

The discussion of the binding of the elements ends with:

> Et hoc est quod dicit Ouidius CONCORDI PACE LIGAVIT et Boetius: Tu numeris elementa ligas.[167]
>
> [And this is what Ovid means by HE BOUND WITH CONCORDANT PEACE, and Boethius by "You bind the elements with numbers."]

And so this brings us back to Boethius, but this time not to the *De institutione arithmetica*, but to to his most famous work, the *De consolatione philosophiae.*

3.9.1.2 Boethius

In the previous quotation, the anonymous commentator on the *Metamorphoses* is referring to a passage in the ninth poem in the third book of the *De consolatione philosophiae* of Boethius:

166 Reading *ex partibus inparibus* for *ex partibus in partibus*. See: Codex Latinus Monacensis 14482 (Munich: Bayerische Staatsbibliothek), f. 28r, accessed December 10, 2025, https://www.digitale-sammlungen.de/en/view/bsb00046312.

167 Böckerman, "Bavarian B, clm 14482c," 340.

> Tu numeris elementa ligas, ut frigora flammis,
> arida conveniant liquidis, ne purior ignis
> evolet aut mersas deducant pondera terras.[168]

> [You bind the elements with numbers, so that cold harmonizes with fire, dry with liquid, lest pure fire be cast away or the weight draw down sunken earth.]

Although set in the context of the *Timaeus*, here Boethius references the Aristotelian contraries (cold/hot and dry/wet) rather than the three contraries of the neoplatonic tradition. This may have contributed to the confusion of the commentator seen above. Moreover, although alluding to the connection between the propotions between numbers and the harmony of the elements, Boethius does not directly mention the relevant geometric sequence itself.

Later commentaries on the *De consolatione philosophiae* connect this passage with the sequence 8, 12, 18, 27, and with the neoplatonic properties of the elements. For example, in his 11th-century commentary on the poem in the ninth chapter of Book 3, Adalbold, bishop of Utrecht from 1010 to 1026,[169] describes the sequence in great detail and indicates their relationship to the three-fold contrary properties of earth and fire.[170] He does not explictly list the properties of all four of the elements, but instead, in a statement similar to that of William of Conches in the *Dragmaticon*, suggests that a diagram might be an aid to understanding:

> Subscribatur figura, ut quibus ad intellectum nostra non sufficit lingua, his ad videndum satisfaciat pictura.[171]

> [A figure is written below so that, for those for whom our words are not sufficient for understanding, they may be satisified through observing the picture.]

This figure, shown in figure 3.19 from the 12th-century manuscript BNF MS Latin 7361,[172] is similar to ones we have seen previously in figures 3.15 and 3.18.

---

168 Boethius, "De consolatione philosophiae," in *De consolatione philosophiae. Opuscula theogica*, ed. Claudio Moreschini (Munich and Leipzig: K. G. Saur, 2005), 3.IX, 80.

169 R. B. C. Huygens, ed., "Serta Mediaevalia: Textus varii saeculorum X-XIII" (Turnhout: Brepols, 2000), 84.

170 Adalbold of Utrecht, "O qui perpetua: Der Kommentar des Adalbold von Utrecht," in *Serta Mediaevalia: Textus varii saeculorum X–XIII*, ed. R. B. C. Huygens (Turnhout: Brepols, 2000), 129–33.

171 Adalbold of Utrecht, "O qui perpetua," 131.

172 "Adalbaldi, Episcopi Trajectensis, opusculum super illud Boëtii: O qui perpetua mundum ratione gubernas," in BNF Latin 7361 (Bibliothèque nationale de France), 46v–56v, accessed December 10, 2025, https://archivesetmanuscrits.bnf.fr/ark:/12148/cc665714.

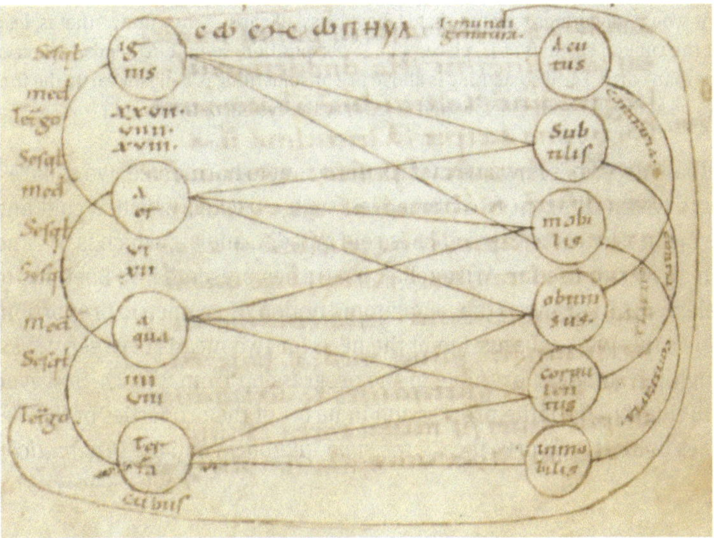

Fig. 3.19: Diagram of the elements in BNF MS Latin 7361, f. 51v

As mentioned earlier, the *Glosae super Boetium*, a commentary on the *De consolatione philosophiae*, was one of the early works of William of Conches. His discussion of the description of the elements in these verses centers first on the Aristotelian contraries:

> Dicatur ergo TV LIGAS id est coniungis elementa terram et cetera. NVMERIS id est proportionibus ita scilicet ut FRIGORA id est terra et aqua, CONVENIANT id est concordent, FLAMMIS scilicit calidis, id est aeri et igni, et ARIDA id est terra et ignis (utrumque enim aridum et siccum est), CONVENIANT LIQVIDIS id est aquae et aeri quorum utrumque humidum et liquidum est. Et ita coniungis, NE PVRIOR IGNIS EVOLET quia, licet naturalis sit motus ignis tendere sursum, ita posuisti eum, ne aliquid esset super ipsum quo tenderet AVT NE PONDERA id est grauitas terrae. DEDVCANT id est deorsum ducant, MERSAS TERRAE id est inferiori loco positas, ne esset aliquid inferius terra quo tenderet. Si enim aliquid esset supra ignem uel aliquid inferius terra, semper illud superius peteret, illud inferius, et ita dissoluerent machinam mundi.[173]

> [Then it is said "you bind," that is, you join together the elements earth and the rest, "with numbers," that is, with proportions, namely such as "cold," that is earth and water, "harmonizes," that is, is concordant with "fire," namely, hot, that is, air and fire, and "arid," that is earth and fire (for both are arid and dry), "harmonizes with liquid," that is, with water and air, which are both wet and liquid. And so you join together "lest pure fire be cast away" because, although it is natural for fire to tend upward, you placed it so there was nothing above

173 Guillelmi de Conches, *Glosae super Boetium*, ed. Lodi Nauta (Turnhout: Brepols, 1999), 166–67.

> it to which it would move "nor weight," that is, heaviness, of earth. "They draw," that is, lead downward, the sunken earth, that is, placed in a lower position, so nothing tends to a place that is lower than earth. For if something is above fire or something is beneath earth, the first must always be above, that other below, and so destroy the system of the world.]

From here William describes how one mean is sufficient to harmonize two squares, or two elements which differ in two properties, such as water, which is cold and wet, being the mean between earth, which is cold and dry, and air, which is hot and wet. He thus manages to use the Aristotelian contraries presented by Boethius to describe the elements in terms of the neoplatonic notion of geometric proportionalities. He follows this with a discussion of the need for two means between cubes, working through the standard example of the progression from 8 to 27. However, he does not provide details of the progression in terms of the properties of the elements as he does in his later works the *Glosae super platonem* and the *Dragmaticon philosophiae*.

3.9.1.3 Biblical References: Wisdom and Genesis

As one would expect in the Middle Ages, discussions of the origin of the elements brought to mind biblical references to the creation of the world. For example, in his commentary on the verses describing the elements in *De consolatione philosophiae*, Adalbold writes:

> *Tu numeris elementa ligas* non adiutorio sed ratione numerorum, quia elementorum ligator numerorum est etiam auctor, quorum exemplo et elementa ligantur et cuncta creantur: *omnia* quippe *in numero et mensura et pondere disposita sunt*, et mensura et pondus sine numero esse non possunt.[174]
>
> ["You bind the elements with numbers," not by the help, but by the reason, of numbers, because he who binds the elements is likewise the author of numbers, by whose example both elements are bound and all things are created: for indeed "all things have been ordered by number, measure, and weight," and measure and weight are not possible without number.]

Here the reference to number, measure, and weight is taken directly from the Book of Wisdom 11:21.[175] Adalbold seems to be saying that the description of the order in nature in Wisdom is consistent with the neoplatonic tenet that relationships between numbers underlie the structure of the world: measure and weight are man-

174 Adalbold of Utrecht, "O qui perpetua," 129–30.

175 *Biblia Sacra Iuxta Vulgatam Versionem*, ed. Robert Weber and Roger Gryson (Stuttgart: Deutsche Bibelgesellschaft, 2007), 1016.

ifestations of numbers, and so number is the ultimate determiner of order in the world.

But, of course, the main creation story in the Bible begins with the Book of Genesis 1:1: *In principio creavit Deus caelum et terram*[176] [In the beginning, God created heaven and earth]. In a Bible, dating from the latter part of the 11th century, from the Saint-Hubert Abbey in Belgium, the decorated initial "IN" from the first word of Genesis includes four circlular bands with the inscriptions

BIS.	BINI.	BIS.	VIII.
BIS.	BINI.	TER.	XII.
TER.	TRIA.	BIS.	XVIII.
TER.	TRIA.	TER.	XXVII.

Inscribed in these bands are personifications of the four elements, ordered fire, air, water, earth.[177] Hence the elements of creation are identified with the neoplatonic geometric sequence 8, 12, 18, 27, in the order we saw in figure 3.12, rather than in the standard order we have seen (see figure 3.20).

3.9.2 Computus

From its origin in the early Middle Ages in reference to calculations of the calendar, in particular in relation to the the date for Easter, *computus* came to refer to any work, including tables and diagrams, dealing with astonomical calculations in the construction of calendars or in the calculation of dates and times more generally. In the broad sense of fitting with studies of the natural world, it is not surprising that *computus* texts might include references to the numerical relationships and harmony between the elements. An interesting example is provided by a figure in a *computus* written by the English monk Byrhtferth.

Byrhtferth was a Benedictine monk who spent most of his life at the Ramsey Abbey near Cambridge. His most famous work, the *Enchiridion*, probably written in the years 1010 to 1012, is a commentary on his own *computus* compilation, which itself dates from the years 988 to 996. This compilation is found, for the most part,

176 *Biblia Sacra Iuxta Vulgatam Versionem*, 6.

177 Harry Bober, "In Principio. Creation Before Time," in *De Artibus Opuscula XL: Essays in Honor of Erwin Panofsky*, ed. Millard Meiss, vol. 1 (New York: New York University Press, 1961), 13–28; Ancien testament: ms. II 1639 (KBR: Bibliothèque royale de Belgique), f. 6v, accessed December 10, 2025, https://opac.kbr.be/LIBRARY/doc/SYRACUSE/10749304; Murdoch, *Album of Science*, 350.

Fig. 3.20: Opening of the Book of Genesis in the Saint-Hubert Abbey Bible

in the 12th-century manuscript St. John's MS 17 in the Bodleian Library at Oxford.[178] The diagram of interest here, found at the end of Byrhtferth's *Epilogus*, which is actually a preface to Bede's *De temporum ratione*, combines all of the aspects of the neoplatonic and Aristotelian descriptions of the four elements that we have seen

178 Peter S. Baker and Michael Lapidge, eds., *Byrhtferth's Enchiridion* (Oxford: Oxford University Press, 1995), xxv–xxiv.

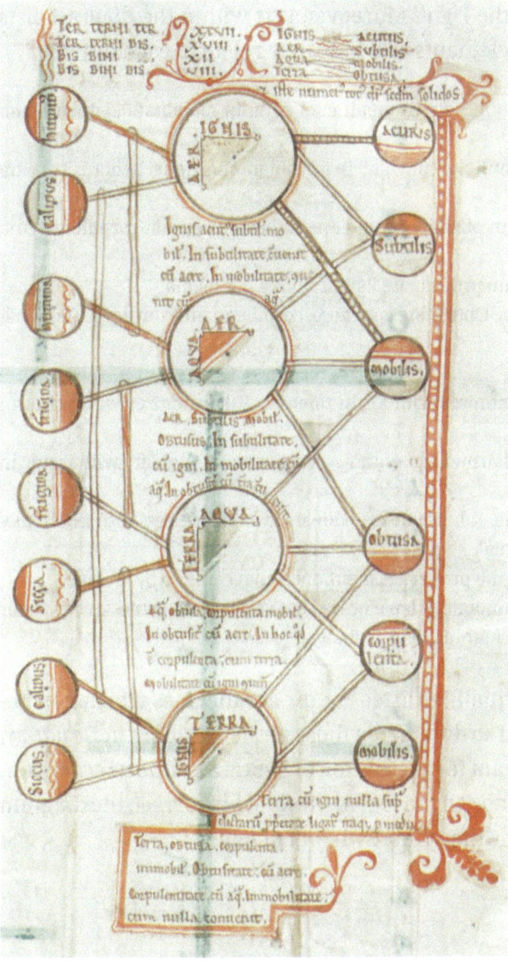

Fig. 3.21: The elements in the *Epilogus* of Byrhtferth

previously.[179] As shown in figure 3.21, the top of the diagram identifies the elements fire, air, water, and earth with the geometric sequence 27, 18, 12, and 8, along with the numbers in factored form. Nearby text states that *iste numerus totus dicitur seculum solidos* [that whole number is called a solid of the world]. The diagram then relates the elements to their properties, with the Aristotelian contraries on the

179 Byrhtferth, "Epilogus," in St John's College MS 17 (Oxford: Bodleian Library, University of Oxford), f. 12v–13r, accessed December 10, 2025, https://digital.bodleian.ox.ac.uk/objects/cca30c56-0751-4f52-a952-bbffcb7b64e9/.

left and the neoplatonic ones on the right. Moreover, text within the diagram lists the neoplatonic properties of the elements and how they relate to each other:

> Ignis acutus, subtilis, mobilis. In subtilitate conuenit cum aere. In mobilitate conuenit cum aqua.
> Aer subtilis, mobilis, obtusus. In subtilitate cum igni. In mobilitate cum aqua. In obtusitus cum terra conuenit.
> Aqua obtusa, corpulenta, mobilis. In obtusitus cum aere. In hoc quod enim corpulenta cum terra, mobilitate cum igni conuenit.
> Terra cum igni nulla supradicatarum proprietate ligatur itaque per media.
> Terra obtusa, corpulenta, immobilis. Obtusitate cum aere, corpulentitate cum aqua. Immobilitate cum nulla conuenit.
>
> [Fire is sharp, fine, mobile. It harmonizes with air in fineness. It harmonizes with water in mobility.
> Air is fine, mobile, and compact. It harmonizes with fire in fineness, in mobility with water, in compactness with earth.
> Water is compact, corporeal, mobile. It harmonizes with air in compactness, with fire in mobility, with earth in that it is corporeal.
> Earth is bound with fire in none of the properties mentioned above and so by means.
> Earth is compact, corporeal, and immobile. It harmonizes with air in compactness, with water in corporeality. It harmonizes with nothing in immobility.]

It is notable that the neoplatonic qualities listed for air are the ones which had become standard since Calcidius, and so differ from those given by Isidore in *De natura rerum* and those listed in the diagram for the edition of Bede's *De natura rerum* published by Hervagius. Moreover, although the text in the figure is correct, the diagram itself connects earth with *mobilis* rather than with *immobilis*.

4 The Geometry of a Circle

Immediately preceding the *Algorismus* in GKS 1812 on folio 13r is a short discussion of various measurements concerning circles, squares, and triangles inscribed in one another. The text was previously edited by Natanael Beckman and Kristian Kålund in *Alfræði Íslenzk*, using AM 415 and AM 685 d as the primary sources,[1] and in *Scriptores rerum danicorum medii ævi*, based on AM 415.[2] The computational algorithms presented here are in the form of simple formulas and, in contrast to the *Algorismus*, are given via examples rather than through general descriptions.

4.1 Geometric Computations

The discussion concerns a diagram (see figure 4.1) displaying a circle with an inscribed square. In turn, another circle is inscribed inside the inscribed square, and, further, an equilateral triangle is inscribed inside the inner circle. As an example of computing the circumference of a circle with a given diameter, the inner circle is taken to have a diameter of 7 ells, from which it follows, using $\frac{22}{7}$ as an approximation for π, that the circle has a circumference of 22 ells. For such a circle, the length of each side of the square circumscribed about it will be 7 ells. And so, for the example, the square will have a circumference of 28 ells.

To find the length of the diagonal of the square, which is also the diameter of the outer circle, the text states:

> Ferſkeyta hver er þriðivngi breidari i ſkacc horn en vm þvert.[3]
>
> [Any square is by a third part broader than around through opposite angles.]

That is, if s is the length of a side of a square and d the length of its diagonal, then

$$2s = d + \frac{1}{3}d = \frac{4}{3}d.$$

From that it would follow that $d = \frac{3}{2}s$, which in this case gives $\frac{21}{2} = 10.5$ ells for the length of the diagonal of the square. This is, of course, only an approximation, the exact value given by $d = \sqrt{2}s$, which is only 9.9 for this example. Hence the calculation given here is equivalent to using $\frac{3}{2}$ as an approximation for $\sqrt{2}$, which follows

1 Beckman and Kålund, *Alfræði Íslenzk*, 231–32.

2 Jacobus Langebek, ed., *Scriptores rerum danicorum medii ævi*, vol. II (Copenhagen: Godiche, 1773), 192.

3 GKS 1812 4to, f. 13r, ll. 5–6.

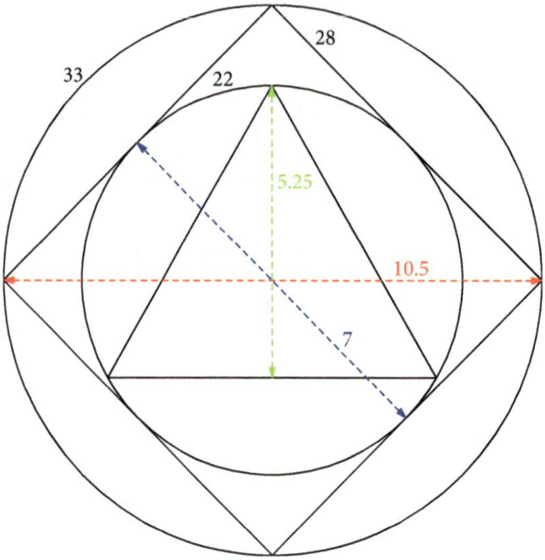

Fig. 4.1: Translated reproduction of the diagram in GKS 1812, f. 13r

from a common rule used in the Middle Ages for approximating square roots.[4] Since $\frac{21}{2}$ is also the diameter of the larger circle, it now follows that the circumference of the larger circle is (approximately)

$$\frac{22}{7} \times \frac{21}{2} = 11 \times 3 = 33 \text{ ells.}$$

This is labeled correctly in the diagram, although written incorrectly as 35 in the text.

The final two computations discussed involve the equilateral triangle inscribed in the inner circle. The first computation notes that the altitude of the inscribed triangle is three-quarters the diameter of the circle, which, in this example, gives an altitude of

$$\frac{3}{4} \times 7 = \frac{21}{4} = 5.25 \text{ ells.}$$

No explanation is provided for this computation, but it follows easily from some basic geometry. Looking at figure 4.2, which shows just the inner circle from the original diagram, we first note that the triangle ABC has a right angle at C and an angle of 30 degrees at A (and hence an angle of 60 degrees at B). It follows that the

4 Namely, to approximate the square root of a number A, starting with a initial approximation a known to be too small (an approximation by *defect*), one may compute an approximation by *excess* as $\sqrt{A} \approx a + \frac{A-a^2}{2a}$. See: David E. Smith, *History of Mathematics*, vol. 2 (New York: Dover, 1958), 254.

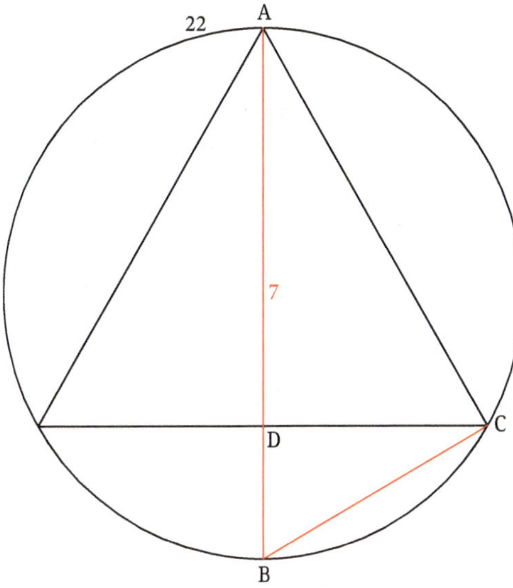

Fig. 4.2: Triangle inscribed in the inner circle

side AC has length

$$7 \times \sin(60^\circ) = 7 \times \frac{\sqrt{3}}{2}.$$

Then triangle ADC has a right angle at D and an angle of 60 degrees at C, and so the length of AD, which is the altitude of the inscribed triangle, is

$$7 \times \frac{\sqrt{3}}{2} \times \frac{\sqrt{3}}{2} = 7 \times \frac{3}{4}.$$

The final result is the least clear. In their edition, Beckman and Kålund indicate that an approximation found in the work of Gerbert of Aurillac is helpful here.[5] In a letter to Adalbold of Utrecht, Gerbert states:

> Et sit tibi regula universalis in omni trigono æquilatero cathetum inveniendi; lateri semper septimam deme, et sex reliquas partes catheto concede.[6]

> [And let the universal rule for finding the altitude of any equilateral triangle be: remove a seventh from the side, and the six remaining parts give to the altitude.]

That is, if s is the length of a side of an equilateral triangle, then the altitude of the triangle is approximately $\frac{6}{7}s$. Equivalently, if h is the altitude of the triangle, one

5 Beckman and Kålund, *Alfrœði Íslenzk*, 232.

6 Gerbert of Aurillac, "Epistola ad Adalboldum," 477.

would approximate the length of a side by $\frac{7}{6}h$. Note that the exact calculation is $h = \frac{\sqrt{3}}{2}s$, so Gerbert's approximation is equivalent to approximating $\frac{\sqrt{3}}{2}$, which is 0.8660 to four decimal places, with $\frac{6}{7}$, which is 0.8571 to four decimal places. With this, one would approximate the length of a side of the equilateral triangle by

$$\frac{7}{6} \times \frac{21}{4} = \frac{49}{8} = 6.125 \text{ ells.}$$

However, the last sentence of the text considers instead an equilateral triangle inscribed in a circle of diameter 6.5:

> En ſida þriſkeytv er halfrar ſettv alnar. ef breidd hringſ er halfrar ſiandv alnar.[7]
>
> [And a side of the triangle is half six ells, if the breadth of the circle is half seven ells.]

Repeating the calculation above for this circle, one would have

$$\frac{3}{4} \times \frac{13}{2} \times \frac{7}{6} = \frac{91}{16} = 5.6875 \text{ ells}$$

for the length of a side of this equilateral triangle. The exact value, to four decimal places, would be 5.6292. However, the text gives 5.5 for the length of the side. It is unclear how the author reached this value, but at least two possibilities suggest themselves. For one, if the altitude of the inscribed triangle was first computed to be

$$\frac{3}{4} \times \frac{13}{2} = \frac{39}{8} = 4.875 \text{ ells,}$$

which might be rounded to 4.75 if the author is working in units of a quarter. A sixth part of 4.875 is about 0.7917, which rounded to quarters would be 0.75. Adding these then gives the reported 5.5. Another possibility would be to treat the side of the triangle as the altitude of an equilateral triangle with sides equal to the diameter of the circle. A direct calculation would give a length of

$$\frac{13}{2} \times \frac{6}{7} = \frac{39}{7} \text{ ells,}$$

which is 5.5714 to four decimal places. Rounded to the nearest quarter, this would be 5.5. Or, since a seventh part of 6.5 is about 0.9286, rounding this to 1 and subtracting from 6.5 also gives a result of 5.5. In any case, given the language used in labeling the lengths of the parts of the diagram, it seems likely that the result of 5.5 comes from some arithmetical calculations being rounded to the nearest quarter.

7 GKS 1812 4to, f. 13r, ll. 11–13.

4.2 Connections with the *Algorismus*

There is a tenuous connection between this text and the *Algorismus*. In the copy in AM 415, the text is followed immediately by the two lines

> Diametr iarðr ııꝶ gradur ok half grada ok eín particula
> Diametr solar ᒣᒣԿ gradur and ok ᒣ *particula*. þat ero ıϴ *dextillor*[8]
>
> [The diameter of the earth is 114 degrees and one-half degree and one particula.
> The diameter of the sun is 225[9] degrees and 2 particulae, that is, 10 dextillor.][10]

The first line expresses the diameter of the earth in terms of multiples of the distance between lines of latitude one degree apart, for which John of Sacrobosco reports an estimate of 700 *stadia*.[11] The result is a simple application of the first computation in the text: Since there are 360 degrees in a full circle, the diameter of the circle is

$$\frac{360}{\frac{22}{7}} = \frac{360 \times 7}{22} = \frac{2520}{22} = 114 + \frac{1}{2} + \frac{1}{22}.$$

The second result, when corrected, follows from the belief at the time that the diameter of the sun was twice that of the earth.[12]

Although these two lines do not appear immediately following the discussion of the geometry of a circle in GKS 1812, they do appear earlier in the manuscript (with the correct value of 229 degrees). On folio 11r, the same two lines, written by the same scribe as the text on folio 13r, appear on either side of the upper half of a diagram illustrating the zodiac. Moreover, these same two statements are found, with the incorrect value for the diameter of the sun, in the diagram illustrating the relationship between the four elements found in AM 736 III that we saw above in figure 3.18. Hence, although GKS 1812 does not connect these statements about the diameters of the earth and the sun with either the *Algorismus* or the geometry of a circle, AMS 415 connects them with the text on the geometry of a circle while AM 736 III connects them with the neoplatonic description of the properties of the four elements.

8 "AM 415 4to," f. 9v, ll. 52–53.

9 This should be 229. The medieval numerals for 5 and 9 are very similar, depending on the scribe's rendering.

10 Although Gustaf Eneström, in his French translation of this passage, renders 'particula' as 'douzième,'—that is, $\frac{1}{12}$—and 'dextillor' as 'minutes,' elsewhere we see 'particula' defined as $\frac{1}{22}$ of a degree and a 'dextill' as $\frac{1}{10}$ of the radius of the sun. See: Langebek, *Scriptores rerum danicorum medii œvi*, 22; Beckman and Kålund, *Alfrœði Íslenzk*, 101–2, 91.

11 Iohannis de Sacro Bosco, "Tractatus de sphera," 83.

12 For example, see: Macrobius, *Commentarii in Ciceronis somnium Scipionis*, I.XX.31, 117.

There is yet another, again tenuous, connection between this text and the *Algorismus*. The value of 700 *stadia*, which John of Sacrobosco gives for the distance between lines of latitude which are one degree apart, is incorrectly recorded in GKS 1812 on folio 20r as 7000 *stadia*. This is notable as this error, copied by the same scribe as the *Algorismus*, appears as well in at least one manuscript of John's *Tractatus de sphera*.[13] Composed sometime in the latter half of the 13th century,[14] this same manuscript, Cambridge University Library MS FF.6.13, includes a copy of an early work of William of Conches, the *De philosophia mundi*.[15] Similar to his later works *Glosae super platonem* and *Dragmaticon philosophiae*, in this text William discusses the properties of the four elements in terms of the neoplatonic contraries and notes the need to join the extreme elements earth and fire by two intermediaries, water and air.[16] However, William does not connect this progression to the geometric progression between the cubes 8 and 27 as he does in his later works.

13 See footnote 79 in Iohannis de Sacro Bosco, "Tractatus de sphera," 85; Cambridge University Library, MS Ff.6.13 (Cambridge: Cambridge University Library), f. 19r.

14 Iohannis de Sacro Bosco, "Tractatus de sphera," 64.

15 Cambridge University Library, MS Ff.6.13, ff. 46r–62r.

16 Cambridge University Library, MS Ff.6.13, f. 49r.

5 The *Algorismus*: Transcript and Translation

The following provides a transcript of the Old Norse *Algorismus* as it appears in the manuscript GKS 1812 4to, folios 13v to 16v. The critical apparatus indicates significant differences in the text found in the manuscripts AM 544 4to (folios 90r to 93r), AM 685 d 4to (folios 24v to 29r), and AM 736 III 4to (folios 4r to 4v). Note that the latter contains only a fragment of the *Algorismus*, starting with *næſt nyian fingr* from line 33 of folio 15r of GKS 1812 and ending with *ef nocur figura ſtendr til hœgri* from line 18 of folio 16r of GKS 1812.

The first modern publication of the Old Norse version of the Indian calculus is P. A. Munch's edition of the version from *Hauksbók* in 1848.[1] Munch included a Danish translation and noted the connection with the *Carmen de algorismo* of Alexander of Villedieu. Finnur and Eiríkur Jónsson edited the version from *Hauksbók* in 1896.[2] Finnur and Eiríkur again identified the text as a paraphrastic translation of the *Carmen de algorismo* and suggested that the original translation might date from before 1270.[3] They based the latter conclusion partially on a comparison of the numerals found in the text with those in a table of medieval numerals compiled by Hermann Hankel,[4] which in turn were derived from a facsimile included in the *Éléments de Paléographie* by Natalis de Wailly.[5] Finnur realized that this comparison was not sufficient to establish the age of the original translation, but considered it nonetheless to be a reasonable supposition.

5.1 The Transcription

Modern transcriptions of medieval manuscripts fall, roughly, into three categories: the facsimile level, which attempts to reproduce the characters, including letters, abbreviation symbols, and related markings and text positioning of the original manuscript as closely as possible for a printed or electronic edition; the diplomatic level, which attempts to reproduce the original while making the text more readable by expanding abbreviations, repositioning text as necessary, and ignoring differences in characters which are not phonologically distinct; and the normalized

1 Munch, "Algorismus."

2 *Hauksbók*.

3 *Hauksbók*, CXXXII.

4 Hermann Hankel, *Zur Geschichte der Mathematik in Alterthum und Mittelalter* (Leipzig: B. G. Teubner, 1874), 325.

5 Natalis de Wailly, *Éléments de paléographie*, vol. 2 (Paris: Imprimerie Royale, 1838), 255–57.

level in which spelling and punctuation are revised into some accepted standard form.[6] These three broad categories are sometimes further divided. For example, the diplomatic level may be divided into transcriptions which expand abbreviations, sometimes indicated by using italics, called the interpretive diplomatic level, and transcriptions that do not.[7]

For Old Norse transcriptions, Guðvarður Már Gunnlaugsson has described a more extensive system of five distinct categories, with each category itself divided into two or three subcategories.[8] What seems most appropriate for a transcription of the Old Norse *Algorismus* in GKS 1812 is that which Guðvarður calls *stafrétt útgáfa*, the second subcategory of his third main category. This is essentially a middle level for diplomatic versions, standing between a more literal version which maintains distinct characters even when they are phonologically indistinguishable, while identifying expanded abbreviations and suspensions in some manner (such as italics and parentheses), and one which ignores differences between phonologically indistinguishable characters and expands abbreviations and suspensions without any identifying marks. This level is substantially the diplomatic level for electronic editions described by the *Medieval Nordic Text Archive*.[9] My intent here is to present a version of the text which, on one hand, clearly shows the distinct differences in style between the text in GKS 1812 and the text from *Hauksbók* in the transcriptions by Finnur and Munch and, on the other, provides a critical apparatus which allows comparison with the other manuscripts.

The transcription provided below of the *Algorismus* from GKS 1812 4to is based on the black-and-white images at the Institut for Nordiske Studier og Sprogvidenskab in Copenhagen which were taken in 1982.[10] Color images are now available at *handrit.is*, a digital library at Landsbókasafn Íslands – Háskólabóksafn in Reykjavík.[11] Many of the pages of the latter have significant dark areas which are largely illegible. The Stofnun Árna Magnússonar in Reykjavík have been very helpful in allowing me to view the manuscript in person and in providing me with updated color photographs which more clearly show the text in the areas obscured in the

6 For example, see: Odd Einar Haugen, "On the Diplomatic Turn in Editorial Philology," in *The Fantastic in Old Norse/Icelandic Literature: Sagas and the British Isles*, ed. John McKinnell, David Ashurst, and Donata Kick (Durham and York: Center for Medieval / Renaissance Studies, 2006), 345.

7 Alfred Foulet and Mary Blakely Speer, *On Editing Old French Texts* (Lawrence, Kansas: The Regents Press of Kansas, 1979), 43–44.

8 Guðvarður Már Gunnlaugsson, "Stafrétt eða samræmt? Um fræðlilegar útgáfur og notendur þeirra," *Gripla* 14 (20003): 197–235.

9 Haraldur Bernharðsson and Haugen, "Levels of Text Representation."

10 GKS 1812 4to.

11 "GKS 1812 4to," in *Handrit.is* (Reykjavík: The Árni Magnússon Institute for Icelandic Studies), accessed December 10, 2025, https://handrit.is/en/manuscript/view/is/GKS04-1812.

older images. Having multiple image sources does create some inconsistencies in readings as, for example, where one source might clearly show an accent another may show none, or where a character appears to be an 'ð' in one image but a 'ꝺ' in another. For the sake of consistency, I have relied primarily on the black-and-white images as they provide the clearest evidence of the text itself.[12] Additionally, I have consulted a partial transcript of the Algorismus in GKS 1812 prepared by Stefán Karlsson, which the editors of *Gripla* previously shared with me.

The influence of Norwegian scribal practices can be seen throughout the text. For example, the use of the 'ꝏ' ligature is seen in such words as *næst* and *bæði*, written as *nꝏſt* and *bꝏðe* in GKS 1812, but as *nęſt* and *będi* in *Hauksbók*. Other Norwegian influences are seen in words such as *hleypr*, written *lỏpir* in GKS 1812, but *hleypr* in *Hauksbók*, as well as words in which 'g' is written 'gh,' such as *merkingh* in GKS 1812 for *merking* in *Hauksbók*. The scribe was not necessarily Norwegian himself, but was likely at least to have been trained in Norway. Haraldur Bernharðsson goes as far as to conjecture that this scribe "may have been trained in the scribal milieu associated with the royal chancery and St. Mary's Church in Oslo in the first half of the fourteenth century."[13] However, the Norwegian influence gives us no indication as to where this particular text may have been transcribed.

The transcript maintains the original punctuation, expands abbreviations, indicates certain types of errors, and distinguishes between the forms of letters only when they might have a phonological difference. In particular, 'f' and 'ꝼ,' 't' and 'ꞇ,' and 'r' and 'ꝛ' are not distinguished. The scribe uses three forms of the letter 'd,' basically in the forms 'ꝺ,' 'δ,' and 'ð.' Although the scribe uses them interchangeably, I have rendered the first two forms as 'd' and the third as 'ð,' even though the text was written at a time when the distinction between the two was disappearing in Icelandic script.[14] The letters 'u' and 'v' are distinguished although the scribe appears to use them interchangeably. Consonant ligatures for 'pp' and 'kk' are expanded silently. The letters 'ſ' and 's' are distinguished. Abbreviation symbols are expanded with italics, small capitals are not expanded, and all accents and punctuation are shown. The 'er/ir' abbreviation for word endings is expanded with 'ir,'

12 For some background on how images were made of the Icelandic manscripts before they were transferred from Denmark back to Iceland, see: Mariane Overgaard, "The Organisation of the Photography and Restoration of Manuscripts Prior to Their Transfer to Iceland 1971–1996," in *Care and Conservation of Manuscripts 3. Proceedings of the Third International Seminar Held at the University of Copenhagen 14th–15th October 1996*, ed. Gillian Fellows-Jensen and Peter Springborg, trans. Gillian Fellows-Jensen (Copenhagen: The Royal Library, 1997), 7–27.

13 Haraldur Bernharðsson, "Scribes and Scribal Practice," 112.

14 For a discussion of the forms of the letter in GKS 1812, see: Haraldur Bernharðsson, "Scribes and Scribal Practice," 104–5, 110–11.

although an argument could be made for either. So, for example, 'fior̉' is expanded as 'fior*ir*,' and 'fing̉r' as 'fing*i*r.' The line breaks are those of the manuscript with the exception that, for the sake of enhancing the readability, words are not broken at the end of lines.

Finally, text added by the editor is indicated as <text> and illegible text as [text]. Words written in the margin, or above or below the line, are not noted as such, but incorporated into the transcript as the scribe intended.

5.2 The Critical Apparatus

The critical apparatus enables comparisons between the four manuscripts. Where two or more manuscripts essentially agree, but differ from GKS 1812, the critical apparatus notes the reading in the oldest manuscript. Minor changes in word order are not noted.

AM 685 d contains a section on mediation not found in the other versions. Beginning at line 20 of folio 25r and continuing through line 15 of folio 25v, and located between the description of subtraction and that of duplation, the section describes mediation in a way similar to that found in the *Algorisums vulgaris* of John of Sacrobosco.[15] In particular, the section describes the representation of $\frac{1}{2}$ as either ♈, a *semis*, or 30 minutes.

The form of the digits 1, 2, 3, 4, 5, 6, 7, 8, 9, and 0 are rendered in close approximation to the form in which they appear in GKS 1812 (see figure 5.1).[16] This is an exception to the general methodology of a diplomatic transcript, but is consistent with the aim of presenting a text which reflects the representation of Indo-Arabic numerals in the 13th and 14th centuries. The form of the digits in the other manuscripts are very similar, with the most variation appearing in the rendering of 5 (see figures 5.2 and 5.3).[17] However, starting with the listing of the even digits (line 21 of folio 13v in GKS 1812), AM 685 d uses a character very close to the modern 2 for ℸ. All differences in the representation of numbers—that is, whether a number is expressed in words, in Roman numerals, or in Indo-Arabic notation—are noted in the critical apparatus.

Variations in spelling, such as Norwegian versus Icelandic usage and changes in vowel usage over time, are not noted. Other minor variations, such as *dupl* and *dufl* and *sik* and *sig*, are passed over as well. Of course, as with word order, what

15 Iohannis de Sacro Bosco, "Algorismus," 181–82.

16 Copenhagen, Det Kongelige Bibliotek, GKS 1812 4to, 13v.

17 Copenhagen, Arnamagnæan Collection, AM 544 4to, 90r, NorS Sprogsamlinger; Copenhagen, Arnamagnæan Collection, AM 685 d 4to, 24v, NorS Sprogsamlinger.

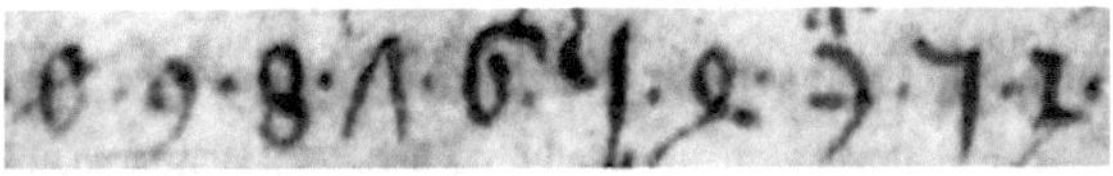

Fig. 5.1: The digits in GKS 1812, f. 13v

Fig. 5.2: The digits in AM 544, f. 90r

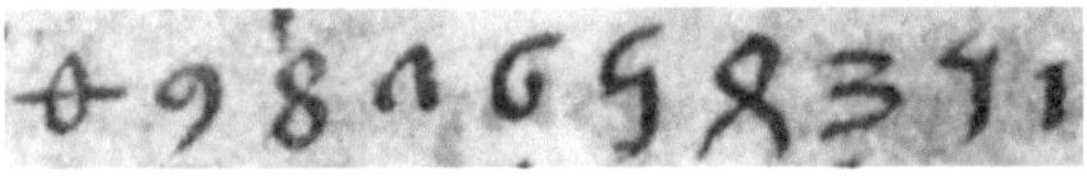

Fig. 5.3: The digits in AM 685 d, f. 24.v

constitutes a minor difference is a subjective judgement. I have tried not to overburden the apparatus with minutia, while still elucidating the ways in which the manuscripts differ.

In the critical apparatus, *A* is AM 544, *B* is AM 685 d, *Γ* is GKS 1812, and *Δ* is AM 736 III.

5.3 The Translation

The translation aims to be as literal as possible while maintaining a smooth flow in English. The intent is to maintain, as much as possible, how the text may have appeared to a reader of that time unfamiliar with the methods of the Indian calculus.

In some places, we have translated *leiða* as "to multiply," which, although not literal, is indicated by the context and seems to be a direct translation of *ducere* in the Latin versions. The text refers to subtraction in two ways, initially as *afdráttr*, a literal translation of the Latin *subtrahere* ("to draw off"), but most frequently as *taka af*. We have translated the latter as "take away"; although it is a less formal way to say "subtract" in English, it is literally correct and maintains the distinction between *taka af* and *afdráttr*. Moreover, it is analogous to the use of the Latin *demere* ("to take away" or "subtract") in the *Carmen de algorismo*.

The *Algorismus* calls the symbol for nothing, ɵ, a *cifra*. John of Sacrobosco notes that this symbol is called by various names:

> Decima vero dicitur theta vel circulus vel cifra, vel figura nihili, quoniam nichil significat.[18]
>
> [But the tenth is called theta or circle or cipher, or figure of nothing, because it signifies nothing.]

Peter of Dacia thought the word *theta* to be derived from the name of an iron used to brand thieves, but it may simply come from its resemblance to the Greek letter θ.[19]

5.4 The manuscripts

5.4.1 GKS 1812 4to (Gammel kongelig samling 1812 quarto)

GKS 1812 4to is a composite manuscript of 36 folios from the 12th to 14th centuries. The inside of the front cover indicates that it was once in the possession of Hákon Ormsson, who was at one time a secretary to Bishop Brynjólfur Sveinsson at Skáholt and later the magistrate of Rangárvallasýsla. Bishop Brynjólfur sent it to the Royal Library in Copenhagen in 1662. The manuscript is now held at the Árni Magnússon Institute in Reykjavík.

The manuscript is primarily the work of four scribes, whose distinct hands serve as the basis for the traditional division of the manuscript into four sections. Sections I and II are the youngest sections, dating from the 14th century. The scribes who wrote these sections were contemporaneous and sometimes collaborated. Section III is from the second half of the 13th century and Section IV is from the late 12th century.

Section IV (ff. 24r–34v) contains a Latin-Icelandic glossary, a computational treatise, a note, in Latin, on the forms of the word vesper, and a short passage from *Íslendingabók* on the calendar and the length of a year.

Section III (ff. 5r–6v, 35r–36v) contains more computational material, a map of the world, a circular diagram with astrological signs along with other labels, a calendar for January and February, and a list of some Icelandic priests.

Section II (ff. 1vb23–40, 13v–18r, 19r13–22r9) contains, along with the *Algorismus*, geographical and astronomical texts, calendrical verses, and a treatment of the division of measurements.

Section I (ff. 1r–4v, 7r–13r, 18v1–19r13, 22r10–23v36) contains, along with the text on the geometry of a circle, *Stjǫrnu-Odda tal*, a diagram of the divisions of philoso-

18 Iohannis de Sacro Bosco, "Algorismus," 176.

19 Smith and Karpinski, *The Hindu Arabic Numerals*, 61; Petrus de Dacia, "Expositio super Algorismum," 88.

phy, diagrams of the planets and the zodiac, illustrations of certain constellations, and more text on computation.

More details on the manuscript may be found in the articles by Haraldur Bernharðsson,[20] Gunnar Harðarson,[21] and Svanhildur Óskarsdóttir and Vasaré Rastonis.[22] Also, see handrit.is (https://handrit.is/manuscript/view/is/GKS04-1812) for further details and a bibliography.

5.4.2 AM 544 4to (Arnamagnæan 544 quarto)

Dating from the late 13th to the middle of the 14th century, AM 544 4to was once, along with AM 371 4to and AM 675 4to, a part of *Hauksbók*. It contains 107 folios on diverse topics written primarily by Haukr Erlendsson as well as Icelandic scribes working for him. The manuscript passed between various hands in the 17th century until eventually passing into the possession of Bishop Brynjólfur Sveinsson at Skáholt in the 1650s. Later Árni Magnússon obtained it from the heirs of Bishop Brynjólfur, and it now resides at the Árni Magnússon Institute in Copenhagen.

The diversity of texts within the manuscript reflects the fact that *Hauksbók* served essentially as a library for Haukr. Along with the *Algorismus*, the text includes works on theology (including an excerpt from the *Elucidarius*, for which a much larger segment is found in AM 675 4to), natural history, and the history of Iceland, along with several sagas. The *Algorismus* itself is positioned between two sagas: *Fóstbrœðra saga* and *Eiríks saga rauða*.

More details on the manuscript may be found in the article on *Hauksbók* by Gunnar Harðarson and Stefán Karlsson.[23] Also, see handrit.is (https://handrit.is/manuscript/view/da/AM04-0544) for more further details and a bibliography.

5.4.3 AM 685 d 4to (Arnamagnæan 685 d quarto)

AM 685 d is a manuscript of 32 folios written in Iceland in the second half of the 15th century. The manuscript is now held by the Árni Magnússon Institute in Copenhagen. Along with the *Algorismus*, it contains the text on the geometry of a circle, as well. In addition, the manuscript has several religious texts, including the *Speculum poenitentis* and a fragment of the *Elucidarius*. The *Algorismus* is preceded by a

20 Haraldur Bernharðsson, "Scribes and Scribal Practice."

21 Gunnar Harðarson, "Medieval Encyclopedic Literature."

22 Svanhildur Óskarsdóttir and Vasareé Rastonis, "Fragments United."

23 Gunnar Harðarson and Karlsson, "Hauksbók."

short religious text and the *De virtutibus et vitiis liber* of Alcuin and it is followed by a short text on measurement and the geometry of a circle.

More details on the manuscript and a bibliography may be found at handrit.is (https://handrit.is/manuscript/view/da/AM04-0685-d).

5.4.4 AM 736 III 4to (Arnamagnæan 736 III quarto)

AM 736 III 4to is a short manuscript of only 4 folios composed in the middle part of the 16th century. In addition to a fragment of the *Algorismus*, the text on the geometry of a circle, and the diagram in figure 3.18, the manuscript contains short passages on motions of the planets, the divisions of Greece, measurement, and the diameter of the earth and the sun. Originally written in Iceland, it now resides at the Árni Magnússon Institute in Copenhagen.

More details on the manuscript and a bibliography may be found at handrit.is (https://handrit.is/manuscript/view/da/AM04-0736-III).

5.5 The *Algorismus* in GKS 1812 4to

f. 13v [L]iſt þeſſi heit*ir* algoriſmus. hana funðo fyſt ínndu*er*ſk*ir* me*n*n m*eð* tíu ſtofu*m* er ſua eru ritadir ·θ·9·8·Λ·ϭ·ч·Ջ·϶·ᒣ·ı· hín fyſtí stafu*ir* m*er*k*ir* eín j fysta ſtað. En annar tua. En þriði. þría. Ok huer eftt*ir* þui sem ſkipad*ir* er alt til hins ſiðarſta er cifra heit*ir*. *ok* ſkaltu þeſſa ſtafuí fra hę́gre hende upp hefuia ok rita till vínſtri hanðar ſem ebreiſku. Hver þeſſi ſtafu*ir* m*er*k*ir* ſik einfalligha í fyrſta ſtað En ef han e*r* j aðrum ſtad þa m*er*k*ir* h*a*n .x. ſinnum ſealfuan ſik. Ok j huern ſtað e*r* þu ſet*ir* nokkora þeſſa figu*r*u. þa m*er*k*ir* hon avalt tíu hlutu*m* meira j þ*eim* ſtað e*r* till uinſtri hanðar veít. helð*ir* en í nœſta ſtad adr. Cifra m*er*k*ir* ekki fir*ir* ſik en hon gerír ſtað. *ok* gefu*ir* adrum figurum merkingh þar nœſt hǿyrir þ*at* till at vita þrenna grein ſtafuanna *ok* allrar tolu. þui at aul tala mínne en tíu. heit*ir* fingi*r*. En ſu tala oll e*r* tegu*m* ge*n*gni*r*. heit*ir* lið*ir*. huart ſem hon e*r* meiri eð*ir* mí*n*ni. En ſu tala e*r* alt e*r* ſaman lið*ir* *ok* fingi*r*. heit*ir* ſamſett tala. Ef þu uilt rita nokkora tolu. þa hyg. þu at. ef þat e*r* fíngi*r*. *ok* ríta j fyſta ſtad eina hu*er*ia figu*r*u ſlika ſem þarf a þeſſa leíð .8. En if þu uilt lid. rita. þa ſettu cifru fir*ir* figu*r*u. a þeſſa lund .Λθ. Vilt þu ſamſetta tolu rita. þa ſettu figu*r*u fir*ir* lid. ſem her .ϭч. hu*er*ia tolu er þu ritar þa er hon íof*n*n ef tígu*m* ge*n*gni*r*. ed*ir* jaf*n*n fíngi*r* er umf*ra*m En oll tala e*r* oiof*n*n. ef oiaf*n*n fingi*r* er umf*ra*m Jafnn*ir* fingi*r* eru fiorír. ᒣ.Ջ.ϭ.8. En oiafnn*ir*. aðrir fior*ir* .϶.ч.Λ.9. En eiɴ er huarki þuí at h*a*n e*r* eigi tala held*ir* uphaf allrar tolu. J. ſeau ſtaði e*r* ſkípt greínu*m* þeſſar liſtar. Heit*ir* hin fyſta uid*ir* laghningh. Annur afdratt*ir* Þriðia tuefaldan Fiorda helmínga ſkiptí. Fimta margfaldan Setta ſkíptíngh. Seaunda at taka rot vndan. oc er ſu m*eð* tueim*ir* greínu*m* Annur e*r* at taka rot vndan fir*ir* ſkǿyttri tolu. En onnur greín er þat at ðragha rot undan atthyrndrí tolu þeirrí e*r* u*er*pilſ uoxt hefu*ir* Fra e*n*ní hǿgre hendi ſkalt þu af taka *ok* vidlegía *ok* ſkipta j helmi*n*ga.

1 m*eð*] ok skipuðu með *A*, *B* **1** tíu] .x. *A* **2** er] þeim er *A*, þeim sem svo *B* **3** En] *om. B* **3** tua] .ij. *A* **3** En] *om. B* **6** ef han er] *.om B* **7** þa] En han er ſkipadr *A*, *om. B* **7** .x.] tíu *B* **8** nokkora] figvrv *A* **8** figu*r*u] annan enn skipat *A* **8** m*er*k*ir*] lyker (?) *B* **8** avalt] jafnan *B* **8** tíu] .x. *A*, *B* **9** veít] er *B* **12** tíu] .x. *A*, *B* **14** lið*ir* *ok* fingi*r*] *om. B* **16** 8] Ջ *B* **17** a þeſſa lund] sem hir er *B* **18** figu*r*u] fingr *A*, *B* **18** her] hir er *B* **21** eiɴ] ı *B* **22** J] *om. B* **24** Fimta] .v. *A* **24** Setta] .vi. *B* **25** ſu] *om. B* **25** m*eð* tueim*ir* greínu*m*] grein a tvęr leidir *A*, *B* **26** er] *om.* 685 d **26** greín] *om. A*, *B* **26** er] *om. B* **27** þat at] *om. B* **27** ðragha] taka *A*, *om. B* **27** rot] *om. B* **28** enní] *om. B*

5.5 The *Algorismus* in GKS 1812 4to

This art is called *algorismus*. First discovered by men of India, they used ten characters, which are so written: 0 9 8 7 6 5 4 3 2 1. The first character denotes one in the first place, the second two, the third three, and each according to how it is placed until the last, which is called a cipher. You shall write these characters from right to left as in Hebrew. **f. 13v**

Each character denotes itself simply in the first place. But if it is in the second place, then it denotes X times itself. And in each place that you place some figure, then it always denotes ten parts more with respect to that place which points to the left, relative to the next place before. The cipher denotes nothing in itself, but it creates a place and gives the previous figures signification.

Next it is appropriate to know a three-fold distinction of the characters and of every number. Every number which is less than ten is called a digit. Every number which is made from groups of ten is called an article, whether it is bigger or smaller. But a number that is both an article and a digit is called a compound number.

If you want to write some number, then examine if it is a digit and write in the first place any figure such as needed, for example, 8. But if you want to write an article, then put a cipher before the figure, like this: 70. If you want to write a compound number, then set the figure before the article, as here: 65.

A number is even if it is made from groups of ten or if an even digit is in the front. A number is odd if the digit in front is odd. There are four even digits: 2, 4, 6, and 8. And the other four are odd digits: 3, 5, 7, and 9. But one is neither[1] because it is not a number but rather the origin of all number.

There are seven branches of this art. The first branch is called addition, the second subtraction, the third doubling, the fourth dividing in half, the fifth multiplication, the sixth division, and the seventh to take a root. And there are two branches for this: One is to take a root of a squared number and another type is to extract a root from an octagonal number which has the shape of a cube.

From the right you should take away from, add, and divide in half. From the left

1 That is, one is neither even nor odd.

En fra uinſtri hendi ſkal þu tuefalda *ok* ſkipta *ok* margfalda Oc ſva dragha vndan rot huaratueggio ☞. Ef þu uilt adra tolu uid adra leggía. þa rita ifu*ir* uppí ena meirí toluna. oc ſet ena mínne tolu iamf*ra*m till e*nn*ar hǒgre handar. Ok leg þa fíguru

f. 14r fyrſt up uid toluna er utarſt e*r* til hǒgre handar oc ef ſu tala oll ſama*n* er fing*ir*. þa rita han j ſama ſtað. En ef talan uerd*ir* ſamſett þa rita j fyrſta ſtad fing*ir*. En legh uidr lidin þa toluna ſem j nœſta ſtad e*r* ad*ir* En ef lid*ir* u*er*d*ir* af uidlagning. j fyrſta ſtad þa rita þar cifru. En legh lidin uid þa tolu er nœſt ſtend*ir* ef þar er nokkur tala ella ríta h*a*n þar ein ſaman. En ef þar er cifra þa tak hana brot. En ſet lidin þar nid*ir* Legh ſidan aðrar fig*ur*ur uid at ſlikum hœtte. ☞ Ef þu uilt adra tolu af annarre taka. þa rita tue*nn*ar tolur. ſem i uid*ir*lagningh. *ok* ſet íamnan ena minne tolu und*ir*. ellighar ia*m*fna. þa tak þu af enne fyrſtu fig*ur*u þa tolu ſem vnd*ir* ſtend*ir* ef þat ma. oc rita ef nokkot e*r* eftt*ir* j ſama ſtad. ella ſet þar cifru. En ef þu mat eighi ena fyrſtu fig*ur*u af taka oc e*r* ſu tala meiri er und*ir* ſtend*ir*. þa tak þu ein af nœſtu fig*ur*u. oc gae þerſſ. at hon geri tio j fyſta ſtadenu*m*. Tak þa af þ*ei*m alla tolu þa ſem und*ir* e*r*. *ok* i ſama ſtad þat ſem af lǒyp*ir*. En ef cifrur ſtanda uppí ifu*ir* þa tak ein af þ*ei*rri fig*ur*u e*r* nœſt ſtend*ir* cifru*m* *ok* rita nio. Þar ſem cifru*m*. voru. alt þar till er þu kem*ir* j þan ſtað e*r* þu uilt af taka. Ok tak þu af þ*ei*m tiu ſem þarf. *ok* rita j ſama ſtad þat er lif*n*nar.[1] Ef þu uilt tuefalða nokkora tolu. þa rita fyrſt ſlika tolu ſem þeer likar. Þar nœſt tuefalda þu þa fig*ur*u er meſt ueit till uinſtri handar *ok* rita j nœſta ſtad þat er af lǒyp*ir* ſua ſem j uid*ir*lagníngh en ef ſemis ſtend*ir* ifu*ir* uppi j yðſta ſtad þa legh uid ein. Þui at þar uar ad*ir* io*m*fn tala e*r* j helmi*n*ga uar ſkift En ef þu uilt helmíngh af taka rita ſlika tolu ſem þu uilt *ok* tak af helm*m*ingh en*ní* fyſtu fig*ur*u ef hon e*r* iofn. En ef hon uar oiofn þa ſkípt j helmí*n*ga þui e*r* af einu*m* lǒyp*ir* *ok* tak vp ein en rita ifu*ir* uppi þan

1 In *B*, this is the end of f. 25r l. 19. The next section begins at f. 25v l. 16.

30 ſva] *om. B* **31** þa] *om. B* **31** uppí] *om. B* **32** tolu] tolu undir *B* **32** ennar] *om. B* **3** ſtad] *om. A* **4** legh] *om. B* **7** at] a *A* **8** i] *om. B* **9** ena] *om. B* **9** þu] *om. B* **9** enne] *om. B* **10** ef] *om. B* **10** ef] *om. B* **10** nokkot] nokkut stendr *B* **10** er] *om. B* **11** cifru] θ *B* **11** ena] j *B* **11** fig*ur*u] ſtad *B* **12** tak þu] takðu *A*, tak *B* **12** oc] *om. A* **13** þerſſ] *om. B* **13** tio] .x. *A* **13** j fyſta ſtadenu*m*] j næſta stad hinum fyrra *B* **13** tolu] taulu *B* **14** *ok*] ok rita *B* **14** ifu*ir*] *om. B* **15** nio] 9 *B* **15** cifru*m*] cifrur *A*, *B* **16** voru] *canc.* ſtoðo **16** tak þu] takðu þar *A*, tak þar *B* **17** þat er] þess er af *B* **19** meſt ueit] *om. B* **20** er] *om. B* **20** ſua] *om. A*, ſe *B* **21** j] *om. B* **21** Þui at] þat *B* **21** io*m*fn] oiofn *B* **23** uar] er *A*, *B* **24** *ok*] en *B* **24** en] ok *B* **24** uppi þan] *om. B*

you should double, divide, multiply, and also extract both types of root.

If you want to add one number to another, then write the larger number above and set the smaller number even to it on the right. Then first add the figure

f. 14r

up to the number which is farthest out to the right. And if this entire number is a digit, then write it in the same place. But if the number is a compound, then write the digit in the first place and add the article to that number which is in the next place before. But if an article results from the addition, then write a cipher in the first place and add the article to that number which stands next if some number is there, or, else, write it there alone. But if there is a cipher there, then remove it and set the article down there. Then add the other figures in the same way.

If you want to take one number away from another number, then write the two numbers as in addition and always set the smaller number below, otherwise even. Then you take away from the first figure the number that stands below if it is possible and, if something is left, write that in same place, or, else, put a cipher there. But if you cannot take away the first figure as that number which stands below is greater, then take one from the next figure and carefully note that this makes ten [added to the figure] in the first place. Then take from this the entire number as is below and [write] what remains in the same place. And if ciphers stand over above, then take one from that figure which stands next to the ciphers and write nine where the ciphers were, all the way until you come to the place where you want to take away from. And you will take from them ten as needed and write what is left in the same place.

If you want to double some number, then first write such number as you like. Next you double that figure which is farthest to the left hand and write in the next place that which remains as in addition. But if semis stands over above in the outermost place, then add one since before there was an even number[2] which was divided in half.

2 This is clearly an error, which appears as well in AM 544 (but not in AM 685 d). The text is referring to the case in which the given number was the result of halving a previous number, resulting in a remainder of one-half. Hence the original number must have been odd (*óiofn*), not even (*iofn*).

ſtaf e*r* helm<i>ngh hue*r*slutar me*r*kir ok uer kollu*m* ſemís *ok* ſua e*r* gior .ᛘ. en ſet cifru j ſtadin. þar nœſt tak af annarre fig*ur*u helmi*n*g at ſama hœtte ef hon er iofn. En ef hon e*r* uiofn. þa tak af helmíngh af þuí e*r* iamt e*r*. oc af up þan *ok* ger af h*onu*m fim j nœſta ſtad þui at þat e*r* helmi*n*g*i*r af tío En ef j adru*m* ſtad ſtendi*r* ein þa tak han up *ok* rita fim j nœſta ſtad. en ſet þar cifru. ſem h*an* ſtoð. Ekki gerir cifra nema nokkur fig*ur*a ſtande till uinſtri handar he*n*ne. Far ſidan fram at ſliku*m* hœtte huerſſu margar ſem fig*ur*ur ero. Ef þer likar at margfalda

f. 14v adra toluna j adra. rita tuennar raðir ſtafuanna m*eð* þ*ei*m hœttí at hín yzſta figura þ*eir*rar tolu er þu marghfaldar ſtandí undír fyſta ſtaf ennar ꝺfre tolu. en til vínſtri handar allar aðrar fra þ*eir*ri þœr ſem und*ir* eru. þar nœſt ſkalt þu hugſa hue*r*ſſu mikít ena meiri fig*ur*u ſkortír a tíu. þa er þu uilt marghfalda. Ok ſua marghar ein<í>*n*gar ſem aſkortír a tíu. ſua opt ſkalt þu ena mínní toluna þaa er þu uilt margfaldda taka af tíghum hennar. *ok* at þu ſkilir þetta margfaldda ſeau *ok* níu. Nío ſkortír eín a tío. þuí tak þu eína ſeau af ſeautighum. þa uerda eftt*ir* þrír *ok* ſextíghí. þ*at* eru ſeau ſínnum níu. At ſlíku ſkapi mat þu aðrar tolur rꝺyna. Margfalda hína fyſtu fig*ur*u retligha j allar þœr er unð*ir* ſtanða. *ok* rita ifu*ir* hue*r*ri fig*ur*u þa margfaldan er hon hefu*ir* oc til uinſtrí handar þat ſem eigí ma ifu*ir* henní í nœſta ſtad m*eð* uidlaghníng rettrí. *ok* þa er þeſſi fig*ur*a er margfolðut fœr ena yzſtu af þ*ei*m ſem und*ir* ſtanda. Vnd*ir* nœſtu fig*ur*u *ok* margfaldda uíd þan ſua ſem vid en*n* fyrra: *ok* ef margfalddan gefu*ir* þer lid ſet cífru ifu*ir* uppí *ok* ſkipa lidnum till vínſtri handar. En ef bœðe u*erð*ir af. margfaldd*ir* fíng*ir* *ok* lið*ir*. þa ríta fíng*ir* ífu*ir* þ*eir*ri fig*ur*u

25 ſtaf] *om. B* 27 En ef hon e*r* uiofn. þa tak af] ella *B* 27 e*r*] at *A*, sem *B* 28 af] hef *B* 28 þan] þa eining *B* 28 h*onu*m fim] henne ƽ *B* 29 tío] .x. *A* 29 fim] .v. *A*, ƽ *B* 30 þar] þar niðr *A* 30 cifra] θ *B* 31 at] *om. B* 32 Ef þer likar at margfalda] En ef þu villt *B* 1 tuennar] tuennar tolur er *B* 2 yzſta] fysta *B* 3 þœr] þar *B* 4 und*ir*] þær *B* 5 a tíu] a .x. *A*, *om. B* 5 Ok ſua marghar ein<í>*n*gar] *om. A*, ok sva marga *B* 6 ſem aſkortír a tíu] *om. A*, sem áá skortir tíu *B* 7 þu] *om. B* 8 ſeau] .víj. *A*, *B* 8 tío] .x. *A* 8 ſeau] .vij. *A*, *B* 9 ſeautighum] .vij. tigum *A*, .Λθ. *B* 9 þrír *ok* ſextíghí] .iij. ok .vi. tigir *A*, .ϭʒ *B* 9 ſeau] .vii: *A*, ʌ *B* 10 níu] 9 *B* 10 mat] ſkaltu *B* 10 Margfalda] eðr því at margfallða *B* 11 j] ok *B* 12 þat] þar *B* 13 m*eð*] j *B* 14 ſem] er *A*, *B* 15 vid] ſem *B* 16 *ok*] enn *A*, *B* 16 En] ſva *B* 17 margfaldd*ir*] *om. A* 17 fíng*ir*] af fingir *B* 17 fig*ur*u] *canc.* talu *Γ*, tolu *B*

But if you want to take half of a number, write such number as you want and take half of the first figure if it is even. But if it was odd, then divide in half that which remains from one less, take up the one and write over above that character which denotes half of any part, which we call semis and make so ♈, and [if it is one, remove it and] put a cipher in that place. Next take half of the second figure in the same way if it is even. But if it is odd, then take half from that which is even and just under that and make from this five, which is half of ten, in the place next to that. But if one stands in the second place, then take it up and write five in next place and put there a cipher where the one stood. A cipher does nothing unless some figure stands to the left of it. Now proceed forward in such way for as many figures as there are.

If you wish to multiply
one number by another, write the two rows of characters in that way that the out- **f. 14v**
ermost figure of that number which you multiply stands under the first character of the upper number, and each of the others of the lower number are to the left. Next you shall consider how much the larger figure which you want to multiply lacks from ten. And as many units as are lacking from ten, that is how often you should take the smaller number, that you want to multiply, from that number of tens. So that you understand this, multiply seven and nine. Nine is one less than ten, therefore you take one seven away from seventy, and then sixty-three remains. That is seven times nine. In the same manner you can attempt other numbers. Multiply the first figure correctly with each figure which stands under and write above each figure the multiple that it has, and to the left that which cannot be over it in the next place using correct addition. Then when this figure is multiplied, move the outermost of those that stand below under the next figure and multiply with that as with the first. And, if multiplication makes it an article, set a cipher over above and arrange the article to the left. But if both a digit and an article result from the multiplication, then write the digit over that figure which you multiplied and the article in the next place. But if only a digit results from the multiplication, then write

er þu margfalddadír. en lid j en*n*œſta ſtað. En ef fíngi*r* eín u*er*di*r* af margf*aldan*. þa rita h*a*n ifui*r* uppí. Ef cifra e*r* i en*n*e ỏfre tolu þa laup ífui*r* hana þuí at ekkí er hennar margfalddan. Þerſſ ſkal en *ok* gá. at taka af figu*r*ur þœr ſem uppí eru ífu*ir* ſettar íamſkíot hu*er*ía ſem þu hefui*r* margfaldat. *ok* rita þan fíngi*r* j ſtad hu*er*rar ſem till hỏyrír edi*r* cífru ef þat er rettarra. En legh þat uid hínar er till vínſtra uegs ſtanda ſem af lỏypi*r*. ef cifra ſtendi*r* ífui*r* þ*ei*rri figu*r*u e*r* þu margfalðar þa tak hana af. ef fíngi*r* u*er*di*r*. oc margf*allda*. ella ſtanðe hon kyr Ef þu grunar huart þu hefui*r* ret margfaldað. þa ſkipt j ſundi*r* alla toluna vm margfalðn. þ*at* er ſu tala er undi*r* ſtod. oc mant þu fa hína ſomu tolu. oc fyrſt hafdír þu. ☞ Ef þu uilt ſkípta j ſundi*r* tolunne þa rita tue*n*nar radír ſtafanna *ok* ríta vndi*r* ena mínne toluna. *ok* ſkal en meírí tala uera h*a*lfu meírí edi*r* þriu ſlik. eði*r* meírí muni*r*. Set þu ena fre*m*rrí figu*r*u þa er undi*r* ſtendi*r* gengt enne fyrſtu. ífui*r* uppí. *ok* aðrar til hỏgre handar iamfram ſem þœr[2] enðaſt er undi*r* ſtanda. Þar nœſt hugſa þu vm hu*er*ſſu opt hin fyſtí fíngi*r* er i oc[3]. einu*m* ỏfra. ſua at íafn opt ſe þœr er fylgía

f. 15r henní hu*er* j þ*ei*rri tolu e*r* ifui*r* ſtendi*r* *ok* ſet þu þan fíngi*r* gengt enne yzſtu figu*r*u er undr ſtendi*r* *ok* þo uppi ífui*r* baðar raðír. Tak ſidan ena fyrſtu af enní fyrſtu figu*r*u *ok* þar nœſt hu*er*ia at hendi íafn opt af e*n*ne ỏfri taulu. En ef ein tala er undi*r* þa tak hana af en*n*e ỏfre[4] tolu*n*ne. Þar nœſt flyt alla tolu þa e*r* undi*r* ſtendi*r* vm eín *ok* fín annan quociens *ok* ſet þan hia hinu*m* fyſta *ok* tak hina neðra tolu ſua opt af enne ỏfre *ok* ger at ſama hœtte ſua opt ſem þarf. Ef þu mat eígí ena nedri[5] tolu edi*r* figu*r*u fínna j enne ỏfre þa ſet þan fíngi*r* er undi*r* ſtendi*r* fremſti*r* nœſt en*n*í fyrſtu *ok* aðrar at ſama hœtti till hỏgre handar *ok* fín ſidan quociens epti*r* ſliku*m* hœtte ok

---

2 The scribe has corrected this from *þer*. 3 The scribe has placed a dot under the preceding *i*, indicating that it should be omitted. However, the text should read *oc i*, as it does in *A* and *B* 4 The scribe has written an extraneous *ennn* after *ỏfre*. 5 The scribe appears to have erased this word, which is in both *A* and *B*.

---

18 j en*n*œſta] ęneſta *A*, j næſta *B* **21** uppí eru ífui*r* ſettar] yfir uppi standa *A*, standa *B* **25** u*er*di*r*] er *B* **26** ſkipt] ſkipt þu *B* **28** fa] þa fa *B* **28** Ef] En ef AM 544 **30** meírí tala] efri talan *B* **30** uera] *om. A, B* **31** fre*m*rrí] *canc.* meiri *Γ*, fremstv *A, B* **33** þœr] þess *B* **33** enðaſt] *om. B* **33** vm] *om. A, B* **34** einu*m*] hinum *B* **34** at] *om. B* **34** opt ſe] skipt sva *B* **1** henní] hinum yzta fingr *B* **1** þ*ei*rri] ſinne *B* **1** ifui*r*] undr *B* **2** yzſtu] fyrſtu *B* **3** opt] oft sem *B* **5** fín] .v. *A* **6** þan] hann *B* **6** hia] hia quociens *B* **6** tolu] tolv .i. qvociens *A* **7** ſem] sem till figuru *B* **8** edi*r* figu*r*u] i figuru *A*, *om. B* **10** ſliku*m*] ſama *B*

it over above. If a cipher is in the upper number, then skip over it because nothing is a multiple of it. And also take care of this, that you remove those figures as are put above as soon as you have multiplied each, and write that digit in the place as belongs to each or a cipher if that is correct, and add that to that which stands to the left side as remains. If a cipher stands over that figure which you multiplied, then remove it if it becomes a digit after multiplication,[3] otherwise leave it standing in place. If you doubt whether you have multiplied correctly, then divide apart the whole multiplied number with that number which stood under. And you will get the same number as you had first.

If you want to divide apart some number, then write the two rows of the characters and write the smaller number underneath. The larger number must be two, three, or more times greater. Set the foremost figure that stands under aligned with the first above and the others to the right, continuing as long as the ones underneath last. Next think about how often the first digit also is in the upper number so that equally often those which follow

it are each in that number which stands over. Set that digit aligned with the outermost figure which stands under and yet above both rows. **f. 15r** After that take the first from the first figure and each after the other equally often from the upper number. But if one number is under, then take it from the upper number. Next move each number that stands under over by one [place] and find another quotient. Set that by the first and take the lower number as often from the upper. Do that the same way as often as needed. If you are not able to find the lower number or figure in the upper, then set the foremost digit which stands underneath next to the first and others in same way to the right and then find the quotient in such a way, and move along the figures as needed. And write each of the quotients together above so many

3 Here we have read *af margfalldan* for *oc margfallda*, which appears to be the more correct sense.

fǒr aftt*ir* fig*ur*ur ſem þarf *ok* ríta alla ſama q*uo*ciens ifu*ir* uppí ſua m*ar*ga ſem þarf. En ef cifra ſtenð*ir* niðri unð*ir* þa laup ifu*ir* hana. þui at ekki ma henni ſkipta. Þa er þu kem*ir* vnd*ir* ena yðſtu fig*ur*u. oc hefu*ir* hinne[6] ſkipt mat þu ekki leng*ir* ſkipta. Oc gœt þu þa þ*eir*rar tolu e*r* eft*ir* ſtend*ir* ef hon e*r* nokkur. Ef þu uilt p*ro*fua huart þu ſkiptír ret þa margfalda þa tolu er und*ir* ſtod uid q*uo*ciens. Oc ma*n*t þu fa ſomu tolu ok fyſt hafðír þu. En ef nokkot líop f*ra*m af j ſkíptíngh þa leg þ*at* uíd ſiðan e*r* margfaldað e*r* *ok* ma*n*t þu fi*n*na hína ſomu tolu Þa er þu leiðir eína hue*r*ía tolu *ok* margfalddar í ſealfua ſik. heit*ir* ſu tala ferſkǒyt ed*ir* quadrans. oc en fyſta tala ſu e*r* þu margfalðar heit*ir* rot. oc e*r* hu*er* tala rot unð*ir* nokkorri[7] tolu En eígí er hu*er* tala ferſkǒyt. Ef þu uilt rot fi*n*na und*ir* nokkorrí tolu. þa ríta fyſt ſlika tolu er þer likar. ok i enu*m* fyſta víofn*n*u*m* ſtad rita und*ir* fing*ir* þan er þu leidír j ſealfuan ſik. ok takí af þat e*r* ifu*ir* uppi e*r* ed*ir* ſua ſem nœſt ma h*a*n ganga. Siðan tuefalða þu þan ſama fíng*ir*. oc heit*ir* þat ðupl. tak þa up fíngrín. oc heít*ir* h*a*n ſubðupl. gœt þu ſubdupls. en ríta ðupl í nœſta ſtad ef þat er fíng*ir*. en ef líð*ir* e*r* þa rita þar ſem fíngrín fyrrí ſtoð *ok* ſet cífru fir*ir*. ella fíng*ir*. ef ſamſet tala. Fín ſiðan nyían fíng*ir* oc <leið>[8] h*a*n j ðupl. oc tak af enne ðfre tolu þa tolu e*r* þu margfaldadír. Sida*n* margfalda þu fíng*ir* j ſealfuan ſik *ok* tak þa tolu af enne ðfre gengt ſealfuu*m* h*onu*m. þar nœſt tuefalda þu fíngrín *ok* gœt ha*nſ* med fyrra ſubduplí. *ok* ſet dupl j nœſta ſtað ſem fyr. Fín þar nœſt[9] nyian fíng*ir* *ok* leid j duplín bœde ſamt. *ok* flyt*ir* duplet fyrra

f. 15v at hinu duplí um eín ſtað. *ok* leg þar uid ef þar ſtod lid*ir* fírír af hinu ðuplinu. Margfalða þa nyian fíng*ir* j bœðe ðuplin. ok tak þa tolu af en*n*e ðfre gengt duplino. Ger at ſama hœtte ſua opt ſem þarf. *ok* leid nyian fíngrín j aul ðuplín. *ok* flyt þau eft*ir* aualt vm eín. þar til er pu kem*ir* j en*n* ydſta ſtad. Ef upp

---

6 The scribe has written *hinnne* here 7 The scribe has corrected this from *nokkoro.* 8 Following Munch. 9 *Δ* begins here.

---

11 fig*ur*ur] fiorv *A* **14** hefu*ir*] *om. B* **14** ſkipt] er *B* **14** þa] þa þa *B* **15** Ef] Enn ef *A* **16** þa] þu þa *B* **16** tolu] *om. B* **17** líop] er *B* **17** j] *om. B* **20** ed*ir*] ok *A*, *om. B* **20** quadrans] *om. B* **20** tala] tald *A* **23** i] ę *A* **25** er] sem *B* **25** ſua] *om. B* **25** ganga] *om. B* **26** ſama] *om. B* **28** þa] þetta þa *B* **28** þar ſem] hann *B* **29** tala] tala er *A*, *B* **29** <leið>] leiðir *B* **32** gœt] *om. B* **32** ha*nſ*] hanſ dufl *B* **33** *ok*] enn *A*, *B* **34** leid] leid þar *B* **34** flyt*ir*] flyt *A*, *B*, *Δ* **34** duplet] dufl hid *B*, hít *Δ* **1** um] j *B* **1** ſtod] var *B* **3** af] *om. B* **4** j] j nyian *B* **4** flyt] *om. Δ* **4** þau] þo *B* **5** enn] *om. Δ*

as needed. But if a cipher stands under, then skip over it because one cannot divide by that. Then when you come under the outermost figure and have divided it, you can no longer divide and then be careful to observe that number which remains, if it is anything. If you want to prove whether you have divided correctly, then multiply that number which stood under with the quotient and you will get the same number that you had at first. And if something remained after the division, then add that afterwards to what is multiplied and you will find the same number.

When you take any number and multiply it by itself, that is called a square, or a *quadrate*, and the first number which you multiply is called a root. Each number is the root of some number, but not every number is a square. If you want to find the root of some number, then first write such number as you like. And in the first place, which is odd, write under it the digit which you multiplied by itself, and take from that which is above or such as comes next to it. After that you double that same digit and that is called a duple. Then record the digit and that is called subduple. Note the subduple and write the duple in the next place if it is a digit, but if it is an article write it where the digit stood formerly and set a cipher before it, or else a digit if it is a compound number. After that find a new digit, multiply it with the duple, and take from the upper number that number which you multiplied. After that multiply the digit by itself and take that number from the upper aligned over it. Then next double the digit, note it together with the former subduple, and set the duple in the next place as before. Then find the next new digit, multiply it with both duples at once, move the former duple

towards the other duple by one place, and add it there if an article remained there **f. 15v**
from that duple. Then multiply a new digit with both the duples and take that number from the upper aligned to the duple. Do this in the same way as often as needed, multiply the new digit with all the duples, and move those always over one until you come to the outermost place. If all of that number is used up which you wrote

geng*ir* oll ſu talan er þu ritadir i fyrſtu*n*ne. þa uar ſu tala ferſkðyt. En rot vnd*ir* þ*eir*ri tolu eru fing*ir* allir ſama*n* þ*eir* er þu tuefalðaðir. m*eð* ſidarſta fíngrinu*m* þ*ei*m e*r* þu fant. Margfalda þu rotína j ſealfua ſik. *ok* mant þu hafua hina ſumu tolu ſem j fyſtu ef þu g*er*dir ret. Ef af lðyp*ir* tolu*n*ne nokkot þa er þu dreghr rotina undan. þa uar ſu tala eígí ferſkðyt *ok* leg þu þa tolu uid hina er þu margfalddír rotina till oc man þu fa ena fyſtu toluna *ok* er ſu tala ol ſama*n* rotín oc af laup rot meiri tolu. En ef fyſti ſtad*ir* þ*eir*rar tolu er þu ritadir uar iafn þa fín fíng*ir* und*ir* nꝏſtu fig*ur*u *ok* margfalða a ſomu leid. ☞ Ef þu margfalðar retligha þa leid ferſkðytta tolu j ſealfua ſik. *ok* ſu tala e*r* af þ*eir*ri margfaldan cem*ir* heit*ir* cubicus ed*ir* u*er*pilſ tala. hun er alla uega iam mikil. En rotin und*ir* cubico uar en ſama. *ok* ferſkðyttrar tolu. huer rot er[10] nokkorrar uerpilſ tolu ed*ir* cubici en eígí er hu*er* tala cubic*us*. Ef þu uilt finna <rot> vndir cubíco. hugſa hu*er*ſſu mikil tala er. *ok* hu*er*ſſu margir ſtadír ero. Fín par nꝏſt fing*ir* j enu*m* fremſta þuſunða ſtað. Þuſunda ſtaði kollum uer þa alla er um þuſundir eínar bríotaſt. þat er en fiordí oc en ſeau<n>dí *ok* en tiundí *ok* en þrettandí ok af afalt lðyp*ir* ifu*ir* tua ſtadi. Fra uinſtrí hendi skal þu þ*e*tta uerk uphefuía. leid þan fíng*ir* er þu fant j ſik cubice þat er tuiſvar ſinnu*m* margfaldad fyrſt j ſealfan ſik oc annat ſín j þa tolu e*r* þar kom af. *ok* þar nꝏſt [ta]k af ðfre tolu þeſſa tolu alla gengt fíng*r*inu*m* þ*ei*m ſealf[um]. Oc þrefalda þar nꝏſt fíngrín. *ok* hoppa ifu*ir* eín ſtad m*eð* þa tolu *ok* ſet j þridía ſtad fir*ir* hínu*m* þa toluna ef þat er fíng*ir*. En ef þat er lid*ir* ſet þar cífru en lidín j nꝏſta

10 A *.b.* written over *rot* and an *.a.* written over *er* indicate that the word order should be inverted.

8 ſidarſta] ſidar *Δ* **9** þu] *om. Δ* **9** mant þu] mantu fa .♈.(?) *B* **9** hina] *om. Δ* **10** ſumu] ſomv *A, B, Δ* **10** ſem] ok *B* **10** ef] ef er *B* **10** tolu*n*ne] af tolunni *B* **12** þu] *om. Δ* **12** þu] *om. B* **12** rotina] rętrnar *A, B* **13** ſu tala] *om. B, Δ* **14** En] *om. A, Δ* **16** þa] aða *A* **16** leid] *om. B* **18** ed*ir*] *om. A, B, Δ* **18** u*er*pilſ tala] *om. B, Δ* **20** rot er] tala er rot *A, B, Δ* **20** uerpilſ tolu] *om. B, Δ* **20** ed*ir*] i *A, om. B, Δ* **20** cubici] kubití tolu *B* **21** <rot>] rot *B, Δ* **23** enu*m*] *om. Δ* **24** um] *om. B* **25** ſeau<n>dí] uijde *B* **25** *ok*] *om. B* **25** tiundí] xdi *B* **25** *ok*] *om. B* **25** af] *om. A, B* **25** afalt] er *B*, íafnann er *Δ* **26** tua] .íj́. *A* **26** ſtadi] ſtadí ꜳ uallt *B* **27** er] *om. A* **28** margfaldad] margfallda þu *B* **31** hoppa] far þa *B* **31** eín] j eín *B* **32** ſtad] *om. Δ* **32** hínu*m* þa toluna] hínne þa þa tolu *B* **33** lid*ir*] liðr þa *A, B, Δ* **33** þar cífru] *Δ* is damaged here.

at the beginning, then that number was a square. And the root of that number is made up of the digits all together which you doubled together with the last digit which you found. Multiply the root by itself and you will have the same number as at first if you have worked correctly. If some of the number remains after you extract the root, then that number was not a square. And you add that number to that which you multiplied the root and you will get the first number. And that number together, the root and remainder, is the root for the larger number. If the first place of that number which you wrote was even, then find a digit under the next figure and multiply in the same way.

If you multiply correctly, then multiply a square number with itself and that number which comes from this multiplication is called a *cubicus*, or cubic, number. It is the same size on all sides. Moreover, the cube root is the same as for the square number: Each number is the root of some cubic number, or *cubicus*, but not every number is a cube.

If you want to find a cube root, note how big the number is and how many places it has. Then find the nearest digit in the farthest of the thousands place. We call places of the thousands all those which break only into thousands. That is the fourth, the seventh, the tenth, the thirteenth, and always jump over two places.

You shall start this work from the left. Multiply that digit which you found with itself cubically. That is twice multiplied, first by itself and a second time by that number which came from there. And next take from the upper number this entire number aligned with that digit itself and next triple the digit, skip over one place with that number, and set that number in the third place before it if it is a digit. But if that is an article, set there a cipher and the article in the next place. And if it is a compound number, set the digit in the same place and the article the next. Next find a new digit in the next place,

ſtad. En ef ſamſet tala er þa ſet fíngrín j ſama ſtad.
en líd et nœſta. Þar nœſt fin nyían fíng*ir* j nœſta ſtad
f. 16r þrefaldri tolu er triple heitir *ok* leid h*an* m*eð* hinni fig*ur*u er fyſt fant þu
Ok uer kollum ſubtripl oc a hǫgra uegh henni j triplit m*eð* margfalðan.
ok þar nœſt leid han ein ſaman i þa tolu er af margfaldan kom
oc uer kollum productum. Tak þa þeſſa tolu alla ſamt af en*n*i ǒfrí
gengt þui er tripl ſtod. þui nœſt le*i*d fing*ir* þan ſama i ſealfuan ſik cubice.
oc tak þa tolu af en*n*i ǒfri gengt ſealfum fingrinum. Tak þan
fing*ir* af *ok* þrefalda her ſem en*n* fyrra. *ok* fin þa nyian fing*ir*. Leið
h*an* m*eð* baðum ſubtriplinn oc triplin ſamt. oc flyt afalt triplin
fornari ept*ir* ſem þu gerir i mi*n*na rotardrat uið dupl. nema her
ſkalt þu afalt ifu*ir* ein ſtad hoppa en leggia þo at ſama hœtte tripl
uid tripl m*eð* rettri uidlagningh Far fram at ſliku hofui meðan
þarf oc þu kem*ir* j yðſta ſtad. En þat skalt þu m*eð* mikilli uanduircd
huxa þa er þu finn*ir* fing*ir*na at þ*eir* taki eigi ſua mikit af ǒfri tolu
at ſu tala hafin eigi ſtað er þu margfaldar triplin till. eð*ir* hín
onnur e*r* þu margfaldar fingrin till þan ſidarra. Varðueit þu afalt
ſubtripl m*eð* tripli. gœt þeſſ ok ef cifrur koma. j subtrípl at engi e*r*
margfaldan ed*ir* þrefaldan þ*eir*ra. en halda þœr ſtoðum ſinu*m* medan nokkur
fig*ur*a er till hǫgre[11] handar þ*ei*m. Ok er þat vuandast j vid*ir*laghníng
tripls. afalt fer þat ſem fyr er rítat j vid*ir*logho liſt. Fíng*ir* allír
ſamt þ*eir* e*r* ſubtripl voro *ok* yztir fing*ir* m*eð* ero rot en*n*ar meirí tolu
þ*eir*rar er þu ritaðir fyſt ef up gek oll talan j af drœttinum. *ok* margfalda
þu ſubtriplin j ſealfuan ſik cubíce. *ok* mun þu finna hína fyrſtu
tolu. En ef aflíop nokkot tolunní j afdrœtti. þa er ſu tala eígi cubíc*us*
En þo er aflaup þat m*eð* ſubtríplum rót nokkurs cubící. Oc ef þu margfaldar
rót ena mínní cubice oc leg uíd þa tolu er af margfaldan
kem*ir* aflaupít. oc mant þu fa fyrſtu tolu er þu rítaðír Oc nu rítum
u*er* at ſinne eigi fleíra þar af. Þeſſar eru fíngra margfaldanar

11 *Δ* ends here.

35 et nœſta] j næſta ſtad *B* **3** ein] i *Δ* **5** er] ſem *B* **5** ſik] ſig i **6** tak] *om. Δ* **6** ǒfri] ef *Δ* **7** Leið] ok leid *Δ* **8** h*an*] þann *A, B, Δ* **8** oc] j *Δ* **8** ſamt] bæði ſamt *Δ* **9** fornari] hinn fornari *B, Δ* **9** i] *om. Δ* **10** ifu*ir*] *om. A* **10** ſtad hoppa] hlaupa *B* **11** rettri] rettr *A* **13** fing*ir*na] figura *B* **14** hafin] hafi *A, B, Δ* **15** onnur] hnígur *Δ* **15** ſidarra] en stærra *B*, hínn siðarra *Δ* **15** þu] *om. Δ* **15** afalt] iafnn *B* **16** e*r*] er hennar *Δ* **17** þ*ei*rra] *om. B, Δ* **17** medan] ef *Δ* **18** er] stendr *B, Δ* **18** vuandast] æ undazt *B* **19** afalt] at avallt *A*, at íafnn *B* **19** fer] er *B* **19** j] *om. B* **20** en*n*ar] enn *A, B* **22** ſubtriplin] triplín *B* **23** þa er] *om. B* **23** eígi] er *B* **25** rót] *om. B* **25** uíd] vndir *A*

treble the number, which is called a triple, and multiply it with the figure which f. 16r you found first, and which we call subtriple, and to the right of it multiply it with the triple. And then multiply it alone with that number which came from the multiplication, which we call the product. Then take this number as a whole from that upper one aligned over where the triple stood. Next multiply that same digit in itself cubically and take that number from that aligned over the digit itself. Take that digit and triple it here as the former and then find a new digit. Multiply it with both the subtriple and the triple together and always move the older triple along as you do in smaller root extraction with duple, except here you shall always skip over one place and add still that same way triple to triple with correct addition. Continue in such way as long as needed and you come to the outermost place. And you shall with great care attend to, when you find the digits, that they not take so much from upper number that that number has no place when you multiply the triple or the other numbers when you multiply the later digit. Always keep subtriple with triple. And note that if ciphers come in a subtriple, nothing is a multiple or triple of them, but they keep to their own places as long as some figure is to the right of them. And what is least difficult is the addition of a triple, so that it always goes as written before in the addition art.

All the digits together, those which were subtriples and the outermost digits too, is the root of the larger number of that which you first wrote if the subtractions used up the whole number. And you multiply the subtriples by themselves cubically and you will find the first number. But if there is some remainder to the number after subtraction, then that number is not a cube. But still the remainder, with the subtriples, forms the root of some cube. And if you multiply the root of the smaller cubically and add the remainder to that number which comes from the multiplication, you can get the first number which you wrote. And now we write at the present no more about this.

ferſkǒyttrar. af .∍. 9. quaðratus. af ˥ ꝶ. Af .ꝶ. ıϬ. quaðratus af .Ϥ ˥Ϥ quadrat*us*. Af .Ϭ. ∍Ϭ. q*ua*dratu*s*. Af .Λ. .ꝶ9. q*ua*dratu*s* Af .8. Ϭꝶ. q*ua*dratus. Af .9. 8ı. q*ua*dratu*s*. Oc er ſu liſt till at finna fíngr*a* margfalðan*ir* ſem ritud er fyr. Þeſſi e*r* fingr*a* margfalddan cubice ∍ rot ˥Λ. cubus .˥ rot 8. cubus .ꝶ. rot Ϭꝶ cubus Ϥ rot .ı˥Ϥ. cubus .Ϭ. rot ˥ıϬ. cubus .Λ. rot .∍ꝶ∍. cub*us* .8. rot Ϥı˥. cubus 9 rot Λ˥9 cubus. Huer ferſkǒyt tala hefu*ir* tuœr mœlínggar. þat er breíd *ok* lengd. En cubicus tala[12] hefu*ir* þre*n*na mœlíngh. þat

f. 16v

er breíd ok lengd *ok* þycd ed*ir* hœd. Oc þui kalla ſpekíngar hu*er*n ſynilighan likama m*eð* þeſſi tolu ſaman ſettan. at h*an* hefu*ir*. jamnan þeſſa mœlíngh þrenna. Med þui at elif ſpeki. oc ein gud uilde heimin fyinlighan *ok* likamlighan ſkapa. þa ſetti han fyrſt tuœr ennar yðſtu hofudſke*m*pnnur eld *ok* jord. þui at ekki ma nat*ur*uligha ſynilight vera uttan þœr Þar sem eldd*ir* gerir lioſ ok rǒringh. En jord ſtaðfeſti ok hald. En m*eð* þui at þau hafua þrenna víam*n*na huiligleika *ok* gangſtadlígha pa. uar nat*ur*uligh naudſyn at ſetia nokkot milli þ*eir*ra þat e*r* ſamþykttí þ*eir*ra vſœtti. Ok ſem fyr er ſaght at eldd*ir ok* jord *ok* þat alt ſem likamlight er er m*eð* þrefaldre tolu er uer kollum cubicum ſaman ſet þa ritum u*er* þeſſa tua cubus. Ritum u*er* jordina þeſſa leid. Tuiſuar ſinnu*m* tueir tuiſuar .˥ꝶ8. En eldin ſua þryſuar þrír þryſ uar .∍9˥Λ. En m*eð* þui at ekki eít mídſkeid ma milli þeſſarra talnna eín*n*a pat er jamre luttegníngh hǒyrír till huartueggia. ok engra annarra tueggia cuba. þa finnu*m* u*er* tuœr lutfellíngar tolur a þeſſa lund. leidum rot enſ meira cubs j quadratum ens mín*n*a[13] cubs þat er tuyſua tueir þryſuar .˥ꝶ. ı˥. Oc rot enſ min*n*a cubs j quadratum ens meira cubs þeím er þryſuar þrír tuyſuar

12 Written above *numeruſſ* in *A*. **13** The scribe appears to have overwritten *meíra* with *mínna*

28 .ꝶ.] *om. B* **29** af] 8 *B* **29** quadratus] *om. A, B* **29** Af] *om. B* **29** q*ua*dratu*s*] af quadratus *A* **30** Af .8. Ϭꝶ] 88 *B* **30** Af] af quadratus af *B* **30** q*ua*dratu*s*] *om. A*, 8 *B* **30** er] *om. A* **32** ˥Λ] 9Λ *B* **33** ˥ıϬ.] .ı.˥.Ϭ. *A* **34** cubus] Rot *B* **34** tuœr] .íj. *A* **35** tala] *om. B* **1** *ok*] *om. B* **1** ed*ir*] ok *B* **1** ſpekíngar] *om B* **2** þeſſi] *om. B* **2** jamnan] avallt *A* **3** elif] eilif *A, B* **4** tuœr] .íj. *A, B* **5** þui at] ſyandi *B* **5** ma] *om. B* **5** uttan þœr] an þeim *B* **8** ſetia nokkot milli] nokkut væri j milli *B* **9** ſem fyr er ſaght at eldd*ir ok* jord *ok* þat alt] med þvi at þat *B* **10** er er] er. þa er. AM 544 **11** u*er*] *om. A, B* **11** þeſſa] a þeſſa *B* **12** tueir] .ij. *A* **12** ˥ꝶ8] *om. B* **12** En] Enn enn *A* **12** ſua] *om. B* **12** þrír] .íij. *A*, ſinnum .3. *B* **13** ∍9˥Λ] *om. B* **13** m*eð*] *Om. B* **13** milli] meðal *A* **14** eín*n*a] fínna *B* **15** ok engra annarra tueggia] *om. B* **15** tuœr] *om. B* **16** leidum] leída *B* **17** tueir] .íj. *A*, .2. *B* **18** j] ok *B* **18** þeím] þat *B* **18** þrír] .3. *B*

These are the digits squared: 3 squared is 9, 2 squared is 4, 4 squared is 16, 5 squared is 25, 6 squared is 36, 7 squared is 49, 8 squared is 64, 9 squared is 81. And the art to find the digits multiplied is as written before. These are the digits cubed: 3 cubed is 27, 2 cubed is 8, 4 cubed is 64, 5 cubed is 125, 6 cubed is 216, 7 cubed is 343, 8 cubed is 512, 9 cubed is 729.

Each square number has two dimensions, that is breadth and length. But a cubic number has three dimensions, that

is breadth, length, and thickness or height. And therefore the sages say every visible **f. 16v** body is put together with this number for it always has these three dimensions.

Owing to eternal wisdom, the one God wished to create a visible and bodily world. Accordingly, he first made two of the outermost elements, fire and earth, because nothing can be naturally visible without them, as fire makes light and movement, but the earth is steadfast and unmoving. And since they have three unequal and opposite qualities, it was then a natural necessity to set something between them which reconciled their differences. And since, as said before, fire and earth, and every such thing that is of a bodily nature is composed out of a three-fold number, which we call cubic, then we write for this two cubes. We write the earth this way: twice times two twice, 2, 4, 8. But fire so: thrice three thrice, 3, 9, 27.

And since in this case no single mean can be between these numbers, with only it being in equal proportion to both and of no other two cubes, then we find two proportional numbers in this manner: We multiply the root of the greater cube with the square of the smaller cube, that is twice two thrice: 2, 4, 12. And the root of the smaller cube with the square of the greater cube, that is thrice three twice: 3, 9, 18. These two numbers belong equally to the outer two cubes since seven and 20 has in itself 18 and half of 18, and 18 has in itself 12 and half of 12. And so has 12 in itself 8

.϶9. ı8. þeſſar tuœr tolur hǒyra íafnt til tueggía hinn*a* enu ydſtu cuba. þuí at ſeau *ok* .xx. hafua j ſer .ı8. *ok* helmíngh af .ı8. En .ı8. hafua j ſer .ı⌉. *ok* helmíngh af .ı⌉. Sua hafua ok .ı⌉. j ſer .8. *ok* helmíng af .8. at ſama hœtti ſkalt þu afalt lutteknýngar fi*n*na millí tueggía cuba. Sua ſkipadí gud tue*n*nar hofutſkepnur millí eldz *ok* íardar. lopt *ok* vatn. oc hefu*ir* uatn tua huiligleika af íord oc tuœr tolur. En af elddí eín huiligleik *ok* eína tolu. En lopt hefu*ir* tua huilighleika af eldí *ok* tuœr tolur. En eín af íorð ok eina tolu. Oc er eldd*ir* þuí lettarí en lopt ſem .⌉Λ. eru meírí en .ı8. En lopt þuí lettarí en uatn ſem .ı8. eru meirí en .ı⌉. vatn þui lettara en jorð ſem .ı⌉. eru meirí en .8. Þetta <er> fullígar at ſkilía j þ*eir*ri fig*ur*u er her er ſiðar gor oc kollud er cubus PErFectus.

---

19 enu] enna *A*, *om. B* **19** ydſtu] yrſtu *A*, yztu *B* **20** ſeau *ok* .xx.] .⌉.Λ. *A*, .2Λ *B* **20** .ı8] atian *B* **20** .ı8.] aian *B* **21** .ı⌉.] .⌉.ı. *A* **21** .ı⌉.] tolf .ı.⌉. *A* **22** afalt] iafnn *B* **24** oc] *om. B* **25** tuœr] .íj *A*, .82. *B* **25** En] *om. A* **25** eín] .ı. *B* **26** tua] .íj. *A*, *om. B* **26** tuœr tolur] .íj. tolur ok tvęr tolur *A* **26** eín] .ꝶ. *B* **27** eldd*ir*] om. *B* **27** eru] er *B* **28** ſem] *om. A* **28** .ı8.] .28. *B* **28** .ı⌉.] .ı9. *B* **29** vatn] vatn er *B* **29** .ı⌉.] .ı9. *B* **29–30** Þetta <er> fullígar at ſkilía] Ma þ<et>ta fvlligar skilia *A*, þetta ma fulluligar ſkílía *B* **30** er her er ſiðar] er ſidar er *B* **30** oc kollud] er kullazt *B*

and half of 8. You shall in the same way always find proportions between two cubes.

Thus God arranged two elements between fire and earth: air and water. And water has two qualities from earth and two numbers, but from fire one quality and one number. And air has two qualities from fire and two numbers, but one from earth and one number. And fire is as much lighter than air as 27 is greater than 18. And air is as much lighter than water as 18 is greater than 12. Water is as much lighter than earth as 12 is greater than 8.

This is to be more fully understood in that figure which is made here later and is called the *cubus perfectus*.

6 The *Carmen de Algorismo*: Transcript and Translation

6.1 The Transcription and Critical Apparatus

The following provides a transcript of the *Carmen de algorismo* of Alexander of Villedieu as it appears in the 13th-century manuscript MS Latin 7475 in the Bibliothèque nationale de France, folios 72r to 78v.[1] As there are numerous manuscripts of the *Carmen de algorismo* with a wide range of variations, no attempt has been made to provide a thorough critical apparatus. The alternative readings which are noted are from the editions of either Halliwell[2] or Steele,[3] or from the manuscript MS. Auct. F. 5. 29 in the Bodleian Library of the University of Oxford.[4] MS Latin 7475 has twenty-eight additional lines beyond the last line of the edition of Halliwell. The first six of these lines appear in both the Steele edition and MS. Auct. F. 5. 29. All twenty-eight of these lines occur in some other manuscripts as well, in particular in manuscript 457 in the collection of Baldassarre Boncompagni[5] and in UB Erfurt, Dep. Erf. CA. 12° 17.[6] The latter is one of the manuscripts used in the edition of Steele, and so will not be used in the critical apparatus except in the final twenty-two lines, not included in that edition. Moreover, alternative readings from the former, as transcribed by Enrico Narducci,[7] will be noted in the final twenty-eight lines.

The aim is to provide a diplomatic transcription following editorial principles similar to those employed in the transcription of the Old Norse *Algorismus*, as described in Section 5.1.

In the critical apparatus, H denotes the edition of Halliwell, Σ the edition of Steele, Θ the manuscript MS. Auct. F. 5. 29, B the manuscript of Boncompagni, and E the Erfurt manuscript.

1 Alexander de Villa Dei, "Carmen de algorismo."

2 Alexander de Villedieu, "Carmen de algorismo."

3 Steele, "Carmen de algorismo."

4 "MS. Auct. F. 5. 29," in *Medieval Manuscripts in Oxford Libraries* (Bodleian Library, University of Oxford), accessed December 10, 2025, https://medieval.bodleian.ox.ac.uk/catalog/manuscript_700.

5 Enrico Narducci, ed., *Catalogo di manoscritti ora posseduti da Baldasssarre Boncompagni*, 2nd ed. (Rome: Tipografia delle Scienze Matematiche e Fisiche, 1892), 271.

6 "Alexander <de Villa Dei>: Carmen de Algorismo," in UB Erfurt, Dep. Erf. CA. 12° 017 (Universitätsbibliothek Erfurt – Handschriften der Amploniana), 68va–70va, accessed September 13, 2023, https://dhb.thulb.uni-jena.de/receive/ufb_cbu_00023580.

7 Narducci, *Catalogo di manoscritti*, 272–73.

The manuscript shows section headings in red, which are here rendered in boldface. Neither *H* nor Σ have section headings; Θ has headings which differ from those in MS Latin 7475. These differences are not noted in the critical apparatus. Common spelling differences are ignored, as well as minor differences in word order or form. In particular, differences such as 'e' for 'ae' or 'æ;' 'v' for 'u';' 't' for 'd' or 'c;' 'i' for 'j,' 'y,' or 'e;' 'n' for 'm;' and 'ſ' for 's' are ignored, as are differences of indicative, subjunctive, and imperative forms of verbs (such as *scribis, scribas*, and *scribe*.) Differences in verb tenses are also ignored.

As with the translation of the Old Norse *Algorismus*, the translation presented here of the *Carmen de algorismo* aims to be as literal as possible while maintaining a smooth flow in English. And, as with the Old Norse *Algorismus* itself, no attempt is made at rendering the translation in verse form.

6.2 Additional Algorithms

The version of the *Carmen de algorismo* presented here concludes with several algorithms not found in the Old Norse *Algorismus*. In six concise lines, the text provides instructions for finding the sum of two types of arithmetical progressions. The first, referred to as a *continua progressio* (a continuous progression—that is, one with successive terms) in the text, is the summation of the first n integers. Although the text breaks the algorithm into two cases, depending on whether the number of terms is odd or even, the sum in both cases is given by

$$1 + 2 + 3 + \cdots + n = \frac{n(n+1)}{2}.$$

The second case, which the text refers to as simply *non continua*, considers the sum of the first n odd or the first n even integers. The formulas given are, for the odd integers,

$$1 + 3 + 5 + \cdots + (2n - 1) = n^2$$

and, for the even integers,

$$2 + 4 + 6 + \cdots + 2n = n(n + 1).$$

After presenting the algorithms for summing those arithmetical progressions, the final section of this version of the *Carmen de algorismo* concludes with twenty-two lines explaining algorithms for five special cases of multiplication. The first repeats an earlier exposition on how to multiply a digit by a digit using the formula

$$a \times b = 10a - a(10 - b),$$

where a and b are integers between 1 and 9 and $a < b$. The other cases consider multiplying an article by an article, an article by a digit, a mixed number (that is, an article plus a digit) by a digit, and, finally, a mixed number by a mixed number.

6.3 The Manuscripts

6.3.1 MS Latin 7475 (Manuscript: Bibliothèque nationale de France Latin 7475)

MS Latin 7475 is a manuscript of 320 folios from the end of the 13th century. It was in the possession of King Charles V in the 14th century, before he gave it to the Dominican convent in Troyes in 1371, from which it passed through two personal libraries before entering the Royal Library in 1732. A treatise on the quadrant precedes the *Carmen de algorismo* and it is followed by the *Algorismus vulgaris* of John of Sacrobosco. The manuscript contains two other texts by John—namely, his *Computus* and the *Tractatus de sphera*—along with two works on minerals, one by Arnold of Saxony and one by Albert the Great.

More details on the manuscript and a bibliography may be found at the Bibliothèque nationale de France (https://archivesetmanuscrits.bnf.fr/ark:/12148/cc94612b).

6.3.2 MS. Auct. F. 5. 29 (Manuscript: Bodleian Auctarium F. 5. 29)

MS. Auct. F. 5. 29 is a composite manuscript in three parts totaling 190 folios from the second half of the 13th century and first half of the 14th. The manuscript was acquired by the Bodleian Library at Oxford University no later that 1602, having previously been given to the College of Saint Mary Winton in 1431. The *Carmen de algorismo* is in the first part (folios 23r to 27v), from the first half of the 14th century, appearing after the *Massa compoti*, also by Alexander of Villedieu, and before some astrological treatises. This part of the manuscript also contains two works by John of Sacrobosco—namely, his *Computus* and the *Tractatus de sphera*—as well as two further tracts on astrology. The second part of the manuscript contains some definitions and propositions from Euclid and the the third part has the first three books of a Latin translation of the *Nicomachean Ethics* of Aristotle.

More details on the manuscript and a bibliography may be found at the Bodleian Library at Oxford (https://medieval.bodleian.ox.ac.uk/catalog/manuscript_700).

6.3.3 Boncompagni 457

Boncompagni 457 is a 13th-century manuscript of 86 folios. Once manuscript 457 in the collection of Baldassarre Boncompagni, it is now MS NAL 657 (Nouvelles acquisitions latines 657)[8] after being purchased by the Bibliothèque nationale de France after in 1898. The *Carmen de algorismo* is the first text in the manuscript, followed by the *Massa compoti*, also by Alexander of Villedieu. Other texts in the manuscript include the *Tractatus de sphera* of John of Sacrobosco, an algorism, and other computus and calendrical material.

For more details on the manuscript, see *Catalogo di manoscritti ora posseduti da Baldasssarre Boncompagni.*[9] Further details and a bibliography may be found at the Bibliothèque nationale de France (https://archivesetmanuscrits.bnf.fr/ark:/12148/cc713666?collect).

6.3.4 UB Erfurt, Dep. Erf. CA. 12° 17 (Universitätsbibliothek Erfurt, Depositum Erfurt Collegium Amplonium 12° 17)

Dep. Erf. CA. 12° 17 is a manuscript of 78 folios from the second half of the 13th century to 1401. Residing in the personal library of Amplonius Rating de Berka in the early 15th century and then donated to Collegium Porta Coeli in Erfurt, it is now held at the Bibliotheca Amploniana at the University of Erfurt. The *Carmen de algorismo* is found in folios 68v to 70v. The text is preceded by the *Massa compoti* (with a few intervening verses on the Zodiac and *computus*), also by Alexander of Villedieu, and followed, after a short religious text, by the Pseudo-Aristotelian text *De signis sive Astronomia (Astrologia) navalis*. Other texts in the manuscript include the *Computus* of John of Sacrobosco and a some verses of *De tabula algoristica*.

More details on the manuscript and a bibliography may be found at the Digitale Historische Bibliothek at the University of Erfurt (https://nbn-resolving.org/urn:nbn:de:urmel-ufb-166747) and at Handschriftenportal (https://www.manuscripta-mediaevalia.de/dokumente/html/obj32205973,T).

8 "Varia de algorismo et computo," in BNF Nouvelles acquisitions latines 657 (Bibliothèque nationale de France), accessed March 1, 2026, https://archivesetmanuscrits.bnf.fr/pageCollections.html?col=1.

9 Narducci, *Catalogo di manoscritti*, 271–73.

6.4 The *Carmen de Algorismo*

f. 72r **Incip*it* prologus**
Hec allorisſm*us* arſ p*re*ſenſ dicit*ur*. in qua.
Talib*us* indo*rum* fruim*ur* biſ quin*que* figu*r*is.
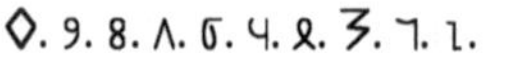
Hic oſt*endit* ſign*ificati*oneſ fig*urarum*[1]
Primaq*ue* ſign*ifica*t un*um*. duo u*er*o ſec*un*da.
Tertia ſign*ifica*t tria. ſic procede ſiniſtre.
Donec ad ex*tr*ema*m* ueniaſ. que cifra uocat*ur*.
It*em* de ſign*ificati*one fig*urarum*[2]
Quelib*et* illaru*m* ſi p*ri*mo limite ponaſ.
Simp*liciter* ſe ſign*ifica*t. ſi uero ſecu*n*do.
Se decieſ. ſurſu*m* p*ro*cedaſ m*u*ltiplica*n*do.
Na*m*q*ue* figu*r*a ſeque*n*ſ q*ue*uiſ ſignat decieſ pl*uſ*.
Ip*s*a locata loco q*ua*m ſign*ifi*cet p*re*eu*n*ti.
Quam p*re*cede*n*teſ pluſ ulti*m*a ſign*ifi*cabit.[3]
Nu*llum* cifra ſign*ifica*t dat ſign*ifi*care ſequenti.
Post p*re*d*i*cta ſciaſ b*re*uite*r*. q*uo*d t*re*s nu*mero*r*um*.
Diſti*n*cte ſpe*cie*ſ ſu*n*t. na*m* quida*m* digiti ſu*n*t.
Articuli quida*m*. q*ui*da*m* quoq*ue* *com*poſiti ſu*n*t.
Sv*n*t digiti nu*me*ri qui cit*ra* denariu*m* ſu*n*t.
Articuli decupli digito*rum*. *com*poſiti ſu*n*t.
Illi qui *con*ſta*n*t ex articuliſ digitiſq*ue*.
Ergo p*ro*poſito nu*mer*o t*ibi* ſc*ri*b*er*e primo.
Reſpiciaſ quiſ ſit nu*m*eruſ. qui ſi digit*us* ſit.
P*ri*mo ſc*ri*be loco ſic. *eſſe* ſ*ed* ſi *com*poſit*us* ſit.
P*ri*mo ſc*ri*be loco digitu*m*. p*o*st artic*u*lum ſic.[4]

---

1 This section heading is in the margin. Neither *H* nor *Σ* have section headings. 2 Section heading is in the margin. 3 *H* omits lines 13-16 and Σ omits line 16, but both add *Quae nil significat; dat significare sequenti.* after line 8. 4 The example .1.3. has been added after this line.

---

2 alloriſm*us*] algorismus *H*, *Σ*, *Θ* **10** Quelib*et*] Quamlibet *Θ* **12** ſurſu*m*] rurſum *Θ* **14** p*re*eu*n*ti] pereunte *Σ* **15** Quam] Nam *Σ* **20** cit*ra* denariu*m*] semper infra decem *H* **24** qui] quia *H*, *Σ*, *Θ* **25** P*ri*mo ſc*ri*be loco ſic] Una figura satis sibi *H*, *Σ*; Primo ſcribe loco digitum *Θ* **25** *eſſe* ſ*ed*] sed *H*, *Σ*; *om.* *Θ* **26** ſic] atque *H*, fac *Σ*, ſit *Θ*

6.4 The *Carmen de Algorismo*

The prologue begins here f. 72r

This art presented here is called algorismus, in which we make such fruitful use of the ten figures of the men of India: 0, 9, 8, 7, 6, 5, 4, 3, 2, 1.

Here he shows the significations of the figures

The first signifies one, the second two, and the third signifies three. So you proceed from the left until you come to the farthest, which is called a cipher.

Further about the signification of the figures

If you place any one of these in the first column singly, it signifies itself. But if in the second column, it signifies ten times itself. Proceed to the higher position by repeating, for each successive figure whatsoever that is placed designates ten times more than it would signify in the preceding place. The last will signify more than the preceding. A cipher signifies nothing; it gives significance to the next figure.

Once you briefly understand the preceding, there are three distinct types of numbers. For some are digits, some articles, and some are compound. Digits are numbers that are less than ten, articles tenfold multiples of digits, and compounds are those which are composed from articles and digits. Hence when a number is proposed for you to write, first you should consider what number it is. If it is a digit, write it in the first place so. But if it is to be a compound, write the digit in the first place behind the article so.[1]

1 The example 13 has been added after this line.

f. 72v
Articulus ſi ſit cifram. post articulum ſit.[5]
Hic dicit quis numeruſ ſit par et quiſ impar[6]
Qvolibet in numero ſi par ſit prima figura.
Par erit et totum quidquid ſibi continuetur.
Impar ſi fuerit. totum ſibi fiet et impar.
Ostendit quot ſunt ſpecies huius artis[7]
Septem ſunt parteſ non pluref iſtiuſ artis.
Addere. ſubtrahere. duplicare que dimidiare.
Sexta eſt diuidere. ſed quinta est multiplicare.
Radicem extrahere. parſ ſeptima dicitur eſſe.
Ostendit in qua parte incipere in quilibet speciem[8]
Svbtrahiſ aut addiſ ad dextriſ uel mediabiſ.
A leua dupla diuide multiplica.
Extrahe radicem duplam ſub parte ſiniſtra.
Hic agit de additione[9]
Addere ſi numerum numero uiſ. ordine tali.
Incipe ſcribe duaſ. primo[10] ſerieſ numerorum.
Primam ſub prima iuſte ponendo figuram.
Et ſic de reliquiſ faciaſ ſi ſint ibi pluref.
Inde duaſ adde primaſ hac conditione.
Si digituſ creſcat ex additione priorum.
Primo ſcribe loco digitum quicumque ſit ille.
Sed ſi compoſituſ in limine ſcribe ſequenti.
Articulum. primo digitum. quia ſic iubet ordo.
Articuluſ ſi ſit in primo limite. cyfram.
Articulum uero reliquiſ inſcribe figuriſ.
Vel per ſe ſcribe ſi nulla figura ſequatur.

5 The example ℓ◊. has been added after this line. **6** Section heading is in the margin. *H* inserts the line *Articulum vero tu in limite scribe sequenti* here; *Σ* inserts the similar line *Articulum vero reliquenti in scribe figure* here. **7** Section heading is in the margin **8** Section heading is in the margin **9** Section heading is in the margin **10** Omitted in *Θ*, but inserted later.

1 post articulum ſit] *H* in primo limite **8** duplicare] duplare *H*, *Σ*, *Θ* **8** que] *om.* *Σ*, *Θ* **9** Sexta] Sextaque *H*, *Θ* **9** eſt] *om.* *Θ* **13** multiplica] multiplicaque *H*, *Σ* **14** duplam] semper *H*, *Σ*, *Θ* **18** Primam] prima *Σ* **18** iuſte] recte *H*, *Σ*, *Θ* **19** ibi] tibi *Σ* **21** priorum] primorum *H* **23** Sed ſi] Si sit *Σ*, Et si *Θ* **23** limine] limite *H*, *Σ*, *Θ* **27** Vel] Et *H*

If it is an article, a cipher is behind the article.[2] **f. 72v**

Here he says which number is even and which is odd

If the first figure of some number is even, the entire number itself will also be even no matter how it is continued. If it is odd, the entire number itself will be odd also.

He shows how many types of this art there are

There are seven parts, not more, of this art: to add, to subtract, to double, and to halve. The sixth is to divide, and the fifth is to multiply. The seventh part is said to be to extract a root.

He shows from which side to begin for each type

You subtract or add from the right, or when you halve. Double, divide, multiply from the left-hand side. Extract a root always[3] from the left side.

Here he speaks about addition

If you wish to add a number to a number, begin with such an order: first write the two rows of numbers, by placing the first figure correctly under the first. And if there are more there, do the same with the remaining. Then add the first two under this condition: If a digit results from the prior addition, write the digit, whatever it may be, in the first place. But if a compound, write the article in the following column,[4] the digit in the first, because the order so requires. If an article, let a cipher be in the first column and write the article with the remaining figures. Or else, if no figure follows, write it by itself.

---

2 The example 40 is has been added after this line.

3 Reading *semper*, as in *H*, Σ, and Θ, for *duplam*.

4 Reading *limite*, as in *H*, Σ, and Θ, for *limine*.

f. 73r Si tibi cif*ra* ſup*er*ueni*en*s occurrerit. illa*m*.
Dele. ſuppo*ſ*itam p*os*t illic ſc*ri*be figura*m*.
Poſtea p*ro*cedaſ reliq*ua*ſ add*en*do figuraſ.

Hic agit de ſubt*ra*ctione

A nu*mero* nu*meru*m ſi ſit t*ibi* demere cura.
Sc*ri*be fig*urarum* ſerieſ ut in addic*ione*.
Maiori nu*mero* nu*meru*m ſuppone minore*m*.
Siu*e* pari nu*mero* ſupponat*ur* nu*meru*ſ par.
Poſt*ea* ſi poſſiſ a p*ri*ma s*u*bt*ra*he p*ri*ma*m*.
Sc*ri*benſ q*uo*d reman*et*. cyfra*m* ſi nil rema*ne*bit.
S*ed* ſi no*n* poſſiſ a p*ri*ma d*e*m*er*e p*ri*ma*m*.
Procedenſ un*um* d*e*limite deme ſeque*n*ti.
Et de*m*ptu*m* p*ro* d*e*nario reputabiſ. ab illo.
S*u*bt*ra*he totale*m* nu*merum* que*m* p*ro*poſuiſti.
Quo f*act*o ſc*ri*be ſup*ra* q*ui*cq*ui*d remanebi*t*.
Fac q*ue* nouenarioſ de[11] cyfriſ d*e*remanebi*t*.
Occurra*n*t ſi forte cyfre. d*e* ſu*m*pſ*er*iſ un*um*.
Poſt*ea* p*ro*cedaſ reliquaſ deme*n*do figuraſ.[12]

Sequit*ur* de duplat*ione*

Si uiſ duplare nu*meru*m. ſic incipe. ſolam.
Sc*ri*be fig*urarum* ſeriem. q*uam*c*umque* uoleſ. et.
Poſtea p*ro*cedaſ p*ri*ma*m* duplica*n*do fig*ur*am.
Inde q*uo*d exc*re*ſcit ſc*ri*baſ ubi iuſſ*er*it ordo.

---

11 The scribe has corrected *deſ* to *de*. **12** *H* and Σ insert four lines here on using addition to verify a correct subtraction.

---

2 Dele] Deme Σ **2** illic] illam Θ **8** ſupponat*ur*] supponatus *H* **9** s*u*bt*ra*he] demere *H* **10** Sc*ri*benſ] Scribas *H* **12** deme] dele Θ **13** Et] Quod *H*, Θ **14** totale*m*] totaliter Σ **15** ſup*ra*] super *H* **16** nouenarioſ] nonenarios *H*, Θ **16** d*e*remanebi*t*] cum remeabis *H*, Θ; cum remanebis Σ **17** d*e* ſu*m*pſ*er*iſ] dum demeris *H*, dum demseris Σ, cum dempſeris Θ **20** ſolam] primo Θ **21** et] tu *H*, Θ; que Σ **22** duplica*n*do] duplando *H*, Σ, Θ **23** exc*re*ſcit] existit *H* **23** ſc*ri*baſ] scribens Σ

If you meet a cipher in the upper row, erase it. Then write the figure that is to be substituted in that place. Afterwards, proceed by adding the remaining figures. f. 73r

Here he speaks about subtraction

If you desire to take away a number from a number, write the row of figures as in addition. Place the smaller number under the larger number, or else equal number is placed under equal number. Afterwards, if possible, subtract the first from the first, writing what remains, a cipher if nothing remains. But if it is not possible to take away the first from the first, proceed to take away one from the the following column and consider that having been taken away as ten from which you subtract the total number which had been set forth. When this is done, write over whatever remains; if ciphers happen to occur, make nines from the ciphers and take one from what remains. Afterwards, proceed to take from the remaining figures.

Next about doubling

If you wish to double a number, begin thus: Write the row of figures alone, however great you wish. And, afterwards, proceed by doubling the first figure. Then write what results where the order designates,

f. 73v Juxta p*re*cepta que dant*ur* in addit*ion*e.
Nam ſi ſit digit*us* in p*ri*mo limite. ſc*ri*be.
Articul*us* ſi ſit in p*ri*mo limite cyfram.
Articul*um* u*er*o reliquiſ in ſc*ri*be figuriſ.
Vel p*er* ſe ſc*ri*baſ ſi nulla figura ſeq*ua*tur.
Compoſit*us* ſi ſit. in limite ſc*ri*be ſeq*ue*nti.
Artic*ulum*. p*ri*mo digitu*m*. quia ſic iub*et* ordo.
Et ſi d*e* reliquiſ facieſ. ſi ſi*n*t ibi plureſ.
Si ſup*er* ex*tr*ema*m* nota ſit monade*m* dat eid*em*.[13]
Q*uo*d tibi *con*tinge*t* ſi p*ri*mo dimidiabiſ.
Atque figu*r*a p*ri*or impar fu*er*it media*n*do.[14]
Sequit*ur* de mediat*io*ne
Incipe ſic ſi uiſ a*liqu*em mu*meru*m mediare.
Sc*ri*be figu*rar*u*m* ſeriem ſolam. uelut an*te*.[15]
Poſt*e*a p*r*o*c*edaſ medianſ et p*ri*ma figura.
Si par. au*t* inpar uideaſ. q*uia* ſi fu*er*it par.
Dimidiabiſ eam. ſc*ri*benſ q*ui*cq*ui*d remanebit.
Impar ſi fu*er*it. unu*m* d*e*maſ. mediare.
Q*uo*d no*n* p*re*ſumaſ. ſ*ed* q*uo*d ſup*er* e*st*. mediabiſ.
Inde ſup*er* tractu*m* fac. de*m*ptu*m* qui notat un*um*.
Si monoſ. dele. ſit *ibi* cyfra. po*ſ*t nota ſup*ra*.
Poſt*e*a p*r*ocedaſ hac. *con*dit*io*ne ſecu*n*da.
Impar ſi fu*er*it. hinc un*um* deme. p*r*iori.

13 Lines 9 through 11 are missing in *H*. **14** Line 11 is missing in Σ. **15** Here, and in lines 18–21, some text has been added above the line.

8 ſi] sic *H*, Σ, Θ **8** ibi] tibi Σ **15** p*r*o*c*edaſ medianſ et p*ri*ma figura] procedens medias, et prima figura Σ; procedas primam mediando figuram Θ **16** au*t*] uel Θ **19** Q*uo*d no*n*] nonne Σ **19** ſup*er* e*st*] superest *H*, Σ, Θ **20** ſup*er* tractu*m*] supertactum *H* **21** ſit *ibi*] cui Θ **23** hinc] hic Σ

just as in the instruction given for addition. For instance, if it is a digit, write it in f. 73v the first column. If it is an article, write a cipher in the the first column and write the article with the remaining figures, or if no figure follows, write by itself. If it is a compound, write the article in the following column, the digit in the first, because the order so requires. And if there are more, do for the remaining in the same manner.[5] If over the end there is a mark, it gives a unit to the same place, which would happen to you if you would have first divided in half and the figure had been odd before having been divided in half.

Next about dividing in half

If you wish to divide some number in half, begin as follows: Write the row of figures alone just as previously. And afterwards proceed to divide the first figure in half. Consider if it is even or odd, because if it should be even, you will divide it in half, writing whatever remains. If it should be odd, you should take away one to divide it in half: not what you had beforehand, but what is left over, divide in half. Then make, over that which was drawn down, that which marks a one. If a one, erase it. Let a cypher be here; subsequently, a mark above. Thereafter, proceed this way with the following provision: If <a figure> is odd, then first take away one,

5 Reading *sic*, as in *H*, Σ, and Θ, for *si*.

f. 74r Infcribenf quinque. nam denof fignificabat.[16]
Monaf predictam. fi uero fecunda det unum.
Illa deleta fcribatur cyfra. priori.
Tradenda quinque pro denario mediato.
Nec cyfra fcribatur nisi deinde figura fequatur.
Poftea procedaf reliquaf mediando figuraf.[17]
Hic agit de multiplicatione
Si tu per numerum numerum uif multiplicare.
Scribe duaf quafcunque uelif ferief numerorum.
Ordo tamen feruetur. ut ultima multiplicantif.
A leua relique fint fcripte multiplicantef.[18]
Ponatur fuper anteriorem multiplicantif.
In digitum curef digitum fi ducere maior.
Per quantum diftet a denif refpice. debef.
Namque fuo decuplo debef delere minorem.
Sic que tibi numeruf uenienf exinde patebit.
Poftea procedaf poftremam multiplicando.
Jufte multiplicanf per cunctaf inferioref.
Conditione tamen tali. quod multiplicantif.
Scribaf in capite quicquid proceſſerit inde.
Sed poftquam fuerit hec multiplicata. figure.
Anteriorentur feriei multiplicantif.
Et fic multiplica uelut iftum multiplicafti.

16 Here, and in the next two lines, some text has been written above the line. There is also additional text in the right margin. 17 *H*, Σ, and Θ have an additional line which elaborates on the previous line. In addition, *H* and Σ have two more lines concerning using doubling to verify a halving. 18 The scribe has indicated that lines 11 and 12 should be switched, as they are in *H*, Σ, and Θ.

2 Monaf] monos *H*, Σ, Θ 2 det unum] sit una *H* 4 Tradenda] Tradendo *H*, Σ, Θ 5 deinde] inde Σ 10 tamen] tum *H* 10 multiplicantif] multiplicandi *H*, Σ, Θ 11 multiplicantef] multiplicantis *H* 14 Per] In Θ 14 quantum] quantes Σ 15 debef] tocies *H*, tociens Σ, Θ 17 poftremam] extremam Θ 18 Jufte] Recte *H*, Θ 19 multiplicantif] mulitplicantem *H*

assigning five, for a one signifies a preceding ten. And if the following is a one, hav- f. 74r
ing erased it, let a cipher be written in the former, five having been traded for half of ten. If a figure does not follow in the next place, a cipher is not written. Afterwards, you should proceed to halve the remaining figures.

Here he speaks about multiplication

If you wish to mulitiply a number by a number, write two rows of whatever numbers you may wish. However, the order is to be kept so the last of the multiplicand[6] is placed over the first of the multiplier, the rest of the multiplier having been written to the left. If you intend to multiply a digit by a digit, consider by how much the larger differs from ten. For indeed you should take away the smaller from ten times itself as many times[7] as follows the number thence coming to be known by you. Thereafter, you should proceed correctly multiplying the last digit by all the lower digits, yet by such condition: You should write on top in that place whatever resulted thence, but after the multiplication of the figure is done, the multiplying row should be brought forward. And so repeat that, just as you multiplied

6 Reading *mutiplicandi*, as in *H*, Σ, and Θ, for *multiplicantiſ*.
7 Reading *tociens*, as in Σ and Θ, for *debeſ*.

f. 74v Qui ſequit*ur* nu*merum*. ſc*r*ipt*um* quiſ*cunque* figuriſ.
S*ed* cu*m* multiplicaſ p*r*imo ſic e*st* ope*r*andu*m*.
Si[19] deb*et* articulum t*ibi* multiplicat*io* ſolu*m*
P*r*epo*ſ*ita cyfra ſumma*m* t*r*anſferre mem*en*to.
Sin au*tem* digit*us* exc*r*eu*er*it articuluſq*ue*.
Articul*us* ſup*rapoſ*ito digito ſalit ult*ra*.
Si digituſ t*amen* poneſ. illu*m*. ſup*er* illam.
Subdita multiplica*n*ſ ha*n*c que ſup*er* in*ſ*id*et* ips*i*.
Deb*et* eam penit*us*. ſc*r*ibenſ q*u*od p*r*oue*n*it in*de*.
S*ed* ſi multipliceſ alia*m*. po*n*ite ſup*er* ip*ſ*am.
Adiu*n*genſ num*erum* que*m* p*r*eb*et* duct*us* earu*m*.
S*i* ſup*ra* inpositam cyfra d*ebet* multiplicare.
Prorſuſ eam debet. ſc*r*ibi q*ue* loco cyf*ra* d*ebet*.[20]
Si fu*er*it ſup*ra* cyfra. ſemp*er* p*r*etereu*n*da.
Si dubitaſ an ſit. bona duplicat*io* f*a*cta.
Diuide totale*m* nu*merum* p*er* multiplicante*m*.
Et reddit nu*mer*uſ em*er*genſ inde p*r*iorem.[21]
Sequit*ur* de d*iuiſione*
Si uiſ diuid*er*e nu*meru*m ſic incipe. p*r*imo.
Sc*r*ibe duaſ q*uaſcunque* ueliſ ſerieſ nu*merorum*.
Maiori nu*mer*o nu*meru*m ſuppone minorem.
Nec dec*et* ut maior teneat biſ t*er*ue m*inor*e*m*.
Et ſub ſupp*r*ima ſupp*r*ima*m* pone figuram.

19 The manuscript has *tibi* after *Si*, a dittography noted as an error by the scribe. 20 *H* and *Σ* have two additional lines here concerning mulitplication by a cipher. The same lines appear in the margin of *Θ*. 21 *Σ* inserts 37 lines on mental multiplication here.

1 Qui] Quæ *H* **1** quiſ*cunque*] quibusque *H*, quicunque *Σ* **3** deb*et*] dabit *H*, *Σ*, *Θ* **4** P*r*epo*ſ*ita] Proposita *H*, *Σ* **5** exc*r*eu*er*it] excrescerit *Σ* **5** articuluſq*ue*] articulusve *H* **7** t*amen*] autem *H* **7** illu*m*] ipsum *H*, *Θ* **7** illam] ipsam *H*, *Σ*, *Θ* **8** in*ſ*id*et*] incidit *H*, *Σ*; incidet *Θ* **8** ips*i*] illi *H*, *Σ*, *Θ* **9** Deb*et*] Delebit *H*; delet *Σ*, *Θ* **9** eam] *om. H* **9** ſc*r*ibenſ] et scribens *H*, scribes *Θ* **9** p*r*oue*n*it] venit *H* **10** ſi] cum *H* **10** multipliceſ] multiplicat *H* **10** alia*m*] illam *Σ* **10** po*n*ite] positam *H*, posite *Σ*, pone *Θ* **11** Adiu*n*genſ] Adjunges *H* **11** p*r*eb*et*] probet *H* **12** ſup*ra* inpositam] supraposita *H* **13** debet] delet *H*, *Σ*, *Θ* **13** q*ue*] quod *H* **13** d*ebet*] dedet *Θ* **14** p*r*etereu*n*da] pretereunda est *H*, *Σ*; est pretereunda *Θ* **15** bona] bene *H*, *Σ*, *Θ* **15** duplicat*io*] multiplicatio *H*, *Θ*; multiplicando *Σ* **15** f*a*cta] secunda *Σ* **17** nu*mer*uſ] numerum *H* **22** Nec] Nam *H*, *Σ*; Sed *Θ* **22** dec*et*] docet *H*, *Σ* **23** ſupp*r*ima] subprima *Θ*

what follows the number by whatever figures have been written. But when you first f. 74v
multiply, it should be done so: If it gives[8] you an article alone, remember to write a cipher to be placed at the end. But if, however, a digit and an article arises, the article jumps beyond by the digit which has been placed above. Yet if it is a digit, you will place it over that. When that which is placed under multiplies that which sits over itself, it erases[9] that completely, writing what resulted in that place. But if you multiply another, place it over itself, adding the number which their multiplication produces. If a cipher ought to multiply that which has been placed above, it erases[10] it completely. A cipher should be written in its place. If a cipher should be above, it should always be passed over. If you are uncertain whether a correct multiplication[11] has been performed, divide the total number by the multiplier and the number appearing thence returns the first.

Next about division

If you wish to divide a number, begin thus: First, write any two rows of numbers you wish. Place the samller number under the larger number. For[12] it is necessary that the larger holds two or three times the the smaller. And place the first figure under the first figure.

---

8 Reading *dabit*, as in *H*, Σ, and Θ, for *debet*.
9 Reading *delet*, as in Σ and Θ, for *debet*.
10 Reading *delet*, as in *H*, Σ, and Θ, for *debet*.
11 Reading *multiplicatio*, as in *Eta* and Θ, for *duplicatio*.
12 Reading *nam*, as in *H* and Σ, for *nec*.

f. 75r Sub reliquiſ reliquaſ ad dext*ra* *par*te locab*is*.
Poſt*ea* d*e* p*ri*ma p*ri*ma*m* ſub *par*te ſiniſt*ra*.
Subt*ra*he ſi poſſiſ quocie*n*ſ poteſ admi*nus*. il*lu*d.
Sc*ri*benſ q*uo*d remane*t*. ſed tali *con*diti*o*ne.
Vt totienſ demaſ. deme*n*daſ a remane*n*ti.
Que ſerie recta ponu*ntur* in i*n*f*er*iori.
Vnica ſi t*amen* ſit ibi. dec*et* hac op*er*ari.
S*ed* ſi no*n* poſſiſ a p*ri*ma dem*er*e p*ri*ma*m*.
P*ro*cedaſ *et* eam nu*mer*o ſuppone ſequenti.
Ha*n*c uno ret*ra*hendo g*ra*du comiteſ ret*ra*hantu*r*.
Et quocienſ pot*er*iſ ab eiſdem deme p*ri*orem.
Vt tocienſ demaſ deme*n*daſ a remane*n*ti.
Nec pluſ q*ua*m nouieſ a*liqua*m t*ibi* dem*er*e debeſ.
Naſcit*ur* hi*n*c nu*mer*uſ quocie*n*ſ. ſup*ra* q*ue* ſeque*n*te*m*.
H*unc* p*ri*mu*m* ſc*ri*baſ. ret*ra*heſ ex in*de* figuraſ.
Du*m* fu*er*it maior. ſup*ra* ſituſ inf*er*iori.
Et rurſuſ fiat diuiſio more p*ri*ori.
Et nu*mer*u*m* quocienſ ſup*ra* ſc*ri*baſ p*re*eunti.
Si fiat ſalt*us* ret*ra*hendo cyfra locet*ur*.
Vt p*re*eat nu*mer*o quocienſ p*re*poniſ eiſdem.
Cyfram. ne nu*mer*i p*er*eat uiſ. du*m* loc*us* illi.
Reſtat. *et* expletiſ diuiſio no*n* u*alet* ult*ra*.
Du*m* fu*er*it nu*mer*uſ minor inf*er*iore ſeorſum.

1 Sub] Sic *H*, *Σ* **3** il*lu*d] istud *Σ* **4** ſed] sub *H*, *Σ* **5** totienſ] toties *H* **6** recta] recte *H*, *Σ* **6** i*n*f*er*iori] anteriori *Σ* **7** t*amen*] tum *H*, tantum *Σ* **7** ibi] tibi *Θ* **7** hac] *om.* *Σ* **10** g*ra*du] gradu quo *Σ* **11** quocienſ] quoties *H* **11** eiſdem] eadem *H*, *Σ*, *Θ* **12** tocienſ] toties *H* **13** a*liqua*m] aliquem *H*, quicquam *Σ* **14** hi*n*c] huic *H* **15** p*ri*mu*m*] primo *H*, *Σ*, *Θ* **15** ret*ra*heſ] retrahens *H*, *Θ* **16** ſup*ra* ſituſ] suppositus *H*, supra positus *Σ*, *Θ* **18** ſc*ri*baſ] scribens *Θ* **18** p*re*eunti] pereunti *H*, *Σ*; preeuntem *Θ* **19** ſalt*us*] saliens *Σ* **20** Vt] et *H*, *Σ* **20** p*re*eat] pereat *H*, *Σ* **20** p*re*poniſ] proponas *H*, *Σ* **20** eiſdem] eidem *H*, *Σ*, *Θ* **21** ne] in *H* **21** nu*mer*i] numerum *H*, *Σ* **21** illi] illic *H*, *Σ* **22** expletiſ] expletus *H* **23** minor] numerorum *Σ* **23** inf*er*iore] in inferiori *Θ*

You will place the rest under the remaining part to the right. Thereafter, if possible, subtract from the first the first under the left part as many times as you can lessen that, writing what remains, but under such condition: As you take away such number of times, those placed in the row below are correctly taken away from the remainder. Even if it is only there once, this ought to be done. But if it is not possible to take away the first from the first, you should proceed and place it under the next number by drawing it, and the associated figures, back by one step. And as often as you are able, take away the former from the preceding; as many times times you should take away that to be taken away from the remainder. You ought to subtract no more than nine times. Henceforth, as often as a number is produced, you should write that above and following the first. Then draw back the figures. While a larger number remains above that situated below, let the division continue in the preceding manner. And you should write the number above as many times as in the preceding. If the step of drawing back results in a cipher located above, place a cipher before the same as precedes the number in the quotient, lest the significance of the number be lost while a place remains for that. And, it having been completed, the division has no further meaning when the number in the higher position is smaller than the one below. **f. 75r**

f. 75v Illu*m* ſeruab*is*. hi*nc* mu*l*tiplicando p*ro*babiſ.
Si ben*e* feciſti diuiſor mu*l*tiplicetur.
Per nu*merum* quocienſ *cum* mu*l*tiplicau*er*iſ adde.
Totali ſu*m*me q*uo*d ſeruatu*m* fuit ante.
Reddet*ur* que t*ibi* nu*mer*uſ q*uem* pr*e*poſuiſti.
Vel ſi nil remane*t* h*unc* mu*l*tiplicat*io* redd*et*.
Se*quitur* de radic*is* ext*ractione* in nu*mer*is *qua*dratis
Cum duciſ nu*merum* p*er* ſe qui p*ro*ue*n*it inde.
Sit t*ibi* *qu*adr*at*uſ. duct*us* radix erit ei*us*.
Nec nu*mer*os om*n*eſ *qu*adr*at*oſ dic*er*e debeſ.[22]
Q*uando* uoleſ nu*mer*i radice*m*. *querere* ſcribi.
Debet dei*nde* noteſ ſi ſit loc*us* ulti*mus* imp*ar*.
Eſt *que* figura loco tal*iſ* ſc*ri*benda ſ*u*b illo.
Que p*er* ſe ducta nu*merum* t*ibi* deſtruit illu*m*.
V*e*l *qu*antu*m* pot*er*it. ex huic delebit eande*m*.
Vel ret*ra*hendo dupleſ ret*ra*heſ dupla*n*do ſ*u*b illa.
Que p*ri*mo se*quitur* apponeſ dei*nde* figuram.
Illi que p*ri*mo duca*tur* p*er* duplicatam.[23]
Poſt p*er* ſe. minue*n*ſ p*ro*poſſe q*uo*d e*st* minue*ndum*.
Hi*nc* illa*m* dele*n*ſ dupla s*u*b ea ſaliendo.[24]
Que ſe*quitur* retr*a*he*n*ſ q*ui*cqu*i*d fu*er*at duplicat*um*.
Poſt hiiſ p*re*poneſ digitu*m* *qui* more p*ri*ori.
P*er* p*re*cedenteſ po*ſ*t p*er* ſe mu*l*tiplicat*us*.

22 This line is followed in *H*, *Σ*, and *Θ* with the line *Est autem omnis numerus radix alicujus.* **23** *H* and *Σ* omit this line. **24** *H* omits lines 20 and 21; *Σ* adds these lines in a footnote.

1 hi*nc*] hunc *H* **2** feciſti] fuisti *H* **5** pr*e*poſuiſti] proposuisti *H*, *Σ*, *Θ* **6** Vel] Et *H*, *Σ* **6** mu*l*tiplicat*io*] multiplicando *Σ* **8** p*ro*ue*n*it] pervenit *H* **9** ei*us*] hujus *Σ* **11** *Quando*] cum *H* **12** dei*nde*] inde *Σ*, ac deinde *Θ* **12** ulti*mus*] ulterius *Σ* **13** illo] ipso *Θ* **14** ducta] dicta *Σ* **15** pot*er*it] poteris *H*, *Θ* **15** huic] inde *Σ*, hinc *Θ* **15** delebit] delebis *H*, *Σ*, *Θ* **15** eande*m*] eadem *H* **16** Vel] Et *H*, Vt *Θ* **16** ret*ra*heſ] retrahens *H*, *Σ*, *Θ* **16** illa] ista *Σ* **17** apponeſ dei*nde* figuram] duplicatur per duplacatam *H*, duplicatur per duplicationem *Σ*, propones deinde figuram *Θ* **18** duca*tur*] duplatur *Θ* **19** p*ro*poſſe q*uo*d e*st* minue*ndum*] sub ea saliendo *H* **20** dele*n*ſ] dele *Σ*, deles *Θ* **20** dupla] duplans *Σ*, duplicans *Θ* **20** ea] ei *Σ* **22** hiiſ] his *H*, *Σ* **22** p*re*poneſ] propones *H*, *Σ*

You will keep that. From this point, you can verify by multiplication: If you have worked correctly, the divisor should be multiplied by the quotient. Once you have multiplied, add to the total sum that which was saved before. And that will restore to you the number number which you had put before. Or if nothing shall remain, the multiplication will restore this. **f. 75v**

Next about the extraction of the root in the case of squared numbers

When you multiply a number by itself, what results thence is for you a square. A root, having been multiplied by itself, will be a square. But you must not say that all numbers are squares.[13] When you want to find the root of a number, it should be written. Then you should observe if the last place is odd. And it is under that place a figure so large should be written which, having been multiplied by itself, removes for you that number, or as much as possible. You should take away[14] the same from this. Additionally, you should double it, drawing back[15] by two places having been retracted beneath those which follow next. You should place the figure there. And let the next figure be multiplied by the double, afterwards by itself, reducing the minuend by the amount possible, hence deleting that double by skipping under it what follows, drawing back whatever had been doubled. Afterwards you should place the digit by those in the way of the preceding manner. In the preceding, after multiplying by itself,

13 *H*, Σ, and Θ add the line "However, every number is the root of some number."
14 Reading *delebis*, as in *H*, Σ, and Θ, for *delebit*.
15 Reading *retrahens*, as in *H*, Σ, and Θ, for *retraheſ*.

f. 76r Deſtruat in q*u*antu*m* pot*er*it un*um* remanen*t*em.
Et ſi p*ro*cedaſ ret*ra*hendo duplica*n*do figura*m*.
P*re*ponen*d*o noua*m* don*ec* totu*m* p*er*agatur.[25]
Si det *com*poſitu*m* nu*merum* duplicat*io*. debet.
In ſc*ri*bi digituſ. ad dextriſ p*ar*te p*ro*pi*n*qua.
Articuloſ q*ue* loco quo ſedit no*n* dupli*ca*ta.
Si dabit artic*ulum*. ſit cyfra loco p*re*eunte.
Articul*us* q*ue* locu*m* ten*et* un*um* dupla*n*do receſſit.
Si donet digitu*m* ſub p*ri*ma pone ſequ*en*te*m*.
Si ſup*ra*po*ſ*ita fiat duplata figura.
Maior. p*re*poni d*ebet* t*antu*m m*odo* cyfra.
Haſ ret*ra*h*e*nſ ſolito p*re*ponaſ more fig*ur*am.
Vſq*ue* ſub ex*tre*maſ ita fac ret*ra*hendo fig*ur*aſ.[26]
Si tot*um* deleſ nu*mer*uſ q*u*em p*re*poſuiſti[27]
Q*ua*drat*us* fuerat de dupla q*uo*d m*u*ltiplicaſti.
Sic q*ue* t*ibi* radix illi*us* certa patebit.
Si d*e* duplatiſ fit iu*n*cta figura ſup*pri*ma.
Radicem p*er* ſe. ſi m*u*ltipliceſ. habeaſq*ue*.
P*ri*mo p*ro*po*ſ*itu*m*. ben*e* te feciſſe p*ro*baſti.
Non e*st* quad*ra*tuſ ſi quid reſtat ſed habe*tur*.
Radix q*ua*drati. qui ſtat maior ſ*u*b eodem.
Et quicq*u*id remane*t* tabula ſeruare mem*en*to.
H*oc* caſu. radix p*er* ſe quo*que* m*u*ltiplicetur.

---

25 After this line, *H* and *Σ* add the line *Subdupla propriis servare docetque duplatis* and Θ the line *Sub dupla sub propriis servare decet duplicatis.* **26** After this line, Θ adds the lines *Eum tamen cifra preponetur duplicate. Est cum sub duplo cifrum leuiare.* **27** After this line, Θ adds the lines *Quadratus fuerat. radicem cius habebis. Sub dupla sub digito iungas sub fine repleto.*

---

1 un*um*] numerum *H*, *Σ*, Θ **2** ſi] sic *H*, *Σ*, Θ **2** p*ro*cedaſ] procedens *Σ* **2** ret*ra*hendo] retrahens *H*, *Σ*, Θ **2** duplica*n*do] duplando *H*, *Σ* **3** P*re*ponen*d*o] Reponendo *H*, Post ponendo Θ **4** duplicat*io*] duplacio *H*, *Σ*, Θ **5** ad] a *H*, *Σ*, Θ **5** dextriſ] dextra *H*, Θ; parte dextra *Σ* **6** Articuloſ] Articulus *H*, *Σ*, Θ **6** ſedit] cedebat *H*, resessit *Σ*, cedit Θ **6** dupli*ca*ta] duplicando *H* **7** p*re*eunte] pereunte *H*, *Σ* **8** dupla*n*do] duplanda *H*, de duplicata *Σ*, duplicanda Θ **8** receſſit] resessit *Σ* **9** ſequ*en*te*m*] sequente *H*, *Σ* **10** fiat] fuerit *H*, *Σ*, Θ **10** duplata] duplicata *H*, *Σ* **11** p*re*poni] proponi *H*, *Σ* **12** Haſ] et Θ **12** p*re*ponaſ] proponas *H*, *Σ* **12** fig*ur*am] figurarum *H* **13** ſub] ad Θ **13** ex*tre*maſ] extrema *H*, *Σ* **14** deleſ] debes *H* **14** nu*mer*uſ] numerum *H*, *Σ*, Θ **14** p*re*poſuiſti] proposuisti *H*, *Σ*, Θ **15** q*uo*d] quem *H*, Θ **15** m*u*ltiplicaſti] duplicasti *H*, *Σ*, Θ **17** fit] sit *H* **18** ſi] *om.* *Σ* **19** feciſſe] fuisse *H* **20** quid] quis *H*, *Σ* **20** habe*tur*] habuentur *H*, habenter *Σ* **21** eodem] eadem *Σ* **22** *Et*] Vel *H*, *Σ* **23** quo*que*] qui Θ

let as much of the remaining number[16] be diminished as it is able. And so[17] proceed f. 76r by retracting doubly, placing a new figure in front until the whole is completed. If doubling produces a compound number, the digit should be written on the near right-hand side and the article sits in that place not doubled. If it shall produce an article, let the cipher be in the preceding place and the article recede one place in the doubling. If it gives a digit, place it next beneath the first. If what is placed over was[18] double, only the larger figure ought to be placed in front of the cipher. Drawing these back, you should place the figure in the usual manner. Do so, drawing back the figures, all the way up to the ends. If you have removed every number,[19] that which had been proposed was a square, from the double that you had multiplied. And so the true root of that will be known, if the combined figure above the first is made from the doubles. If you multiply the root by itself and you get the number first proposed, you are demonstrated to have done well. It is not a square if something remains, but a square root is found which is greater beneath the same place. And remember to save whatever remains on the counting board. Also, in this case the root is multiplied by itself

16 Reading *numerum*, as in *H*, Σ, and Θ, for *unum*.
17 Reading *sic*, as in *H*, Σ, and Θ, for *si*.
18 Reading *fuerit*, as in *H*, Σ, and Θ, for *fiat*.
19 Reading *numerum*, as in *H*, Σ, and Θ, for *numerus*.

f. 76v *Et* ſi quadratus ſub primo maior habetur.
Huic addenſ remanenſ et primus debet haberi.
Si locus extremus fuerit par. ſcribe figuram.
Sub preeunte loco per quam debeſ operari.
Que quantum poterit ſupprimaſ deſtruit ambaſ.
Et penitus legem teneaſ operando priorem.
Si ſupra poſitam digituſ sub fine repertus.
Omnino delet illic ſcribi cyfra debet.
A leua ſi qua ei ſit ſociata figura.
Si cyfre remanent in fine pareſ decet harum.
Radiciſ numerum mediam preponere partem.
Tali queſita radix patet arte reperta.
Sequitur deradicis extractione in numeriſ cubiciſ
Per numerum recte ſi noſti mutiplicare.
Eius quadratum. numeruſ qui prouenit inde.
Dicetur cubicus primus radix erit eiuſ.
Nec numeroſ omneſ cubicatoſ dicere debeſ.
Est autem numeruſ omniſ radix cubicati.
Si cureſ cubici radicem querere primo.
In ſcriptum numerum diſtinguere per loca debeſ.
Que tibi mille notant a mille notante suprema.
Inicium ſumaſ operandi parte ſiniſtra.
Illic inſcribeſ digitum qui multiplicatus.

1 *Et*] Vel *Σ* **1** ſi] sic *H, Σ, Θ* **2** Huic] Hinc *Σ* **2** addenſ] addas *H, Σ, Θ* **2** primus] prius *H, Σ* **2** debet] debes *H, Σ* **2** haberi] habere *H* **4** preeunte] pereunte *H, Σ* **6** Et] Vel *H, Σ* **7** ſupra poſitam] suppositum *H, Σ* **7** sub] in *H*, suo *Σ* **11** Radiciſ] Radices *H, Σ* **11** numerum] numero *H, Σ, Θ* **11** preponere] propone *H*, proponere *Σ* **15** prouenit] pervenit *H, Σ* **18** cubicati] alicujus *H, Σ* **21** suprema] supprima *H, Θ* **22** Inicium] Junctum *H*, Initiam *Σ* **22** ſumaſ] summes *H*, summa *Σ* **23** inſcribeſ] et scribas *H*; sub scribas *Σ*; subscribas *Θ*

and if the square under the first is found greater, you should add[20] what remains to this and the first ought to be found. If the furthest place is even, write a figure under the preceding place by which you ought to produce so much that it will be able to eliminate the first two above. And you should maintain the method of operating entirely as before. If a digit having been placed above at the end is found to be removed entirely, a cipher out to be written in that place. If there is anything to the left, the figure should be joined to it. If ciphers remain at the end, it is correct to place half the number of the pairs of them in the front part of the root. In such way the sought for root is shown to have been found by this method. **f. 76v**

Next about the extraction of the root in the case of cubic numbers

If you have learned to multiply correctly a number by its square, the number which thereupon results is called a cube. One is the root of itself. And you ought not to say that all numbers are cubes. However, every number is the root of a cube. If you desire to find the root of a cube, first you ought to separate the number which has been written by places, those designating a thousand are noted to you with a mark, beginning to end by a thousand. You should begin working from the left side. There you will write a digit which, multiplied

20 Reading *addas*, as in *H*, Σ, and Θ, for *addenſ*.

f. 77r In ſemet cubice ſup*ra* positu*m* t*ibi* p*er*dat.
Et ſi quid fu*er*it adiu*n*ctu*m* p*ar*te ſiniſt*ra*.
Si no*n* o*mnino* q*ua*ntu*m* pot*er*it minue*n*do.
Hi*nc* t*ri*plica*n*ſ ret*ra*he ſaltu*m* facie*n*do ſ*u*b illa.
Q*ue* mane*t* a digito deleto t*er*na. figura*m*.
Illi p*re*poneſ q*ue* ſub t*ri*plo ſociet*ur*.
Vt cu*m* ſub t*ri*plo p*er* eam t*ri*pla m*u*ltiplice*tur*.
Hi*nc* p*er* eam ſolam p*ro*ductu*m* m*u*ltiplicab*is*.
Poſtea totale*m* nu*merum* qui p*ro*uenit in*de*.
A ſup*ra* positiſ reſp*e*ctu pone t*ri*plate.
Addita ſup*pri*mo cubice tu*nc* m*u*ltiplice*tur*.
Reſp*e*ctu cuiu*ſ* nu*mer*uſ qui p*rogre*dietur.
Ex cubico ductu ſup*ra* poſitiſ adimet*ur*[28]
Tu*nc* ip*ſ*am dele*n*ſ t*ri*pleſ ſaltu*m* facie*n*do.
Semp*er* ſub t*er*naſ ret*ra*haſ aliaſ t*ri*plicataſ.
Ex hi*nc* t*ri*platiſ aliam p*re*pone figura*m*.
Que p*er* t*ri*plataſ ducat*ur* more p*ri*ori.
P*ri*mo ſub t*ri*pliciſ subiu*n*ctiſ poſtea p*er* ſe.
In nu*merum* ducta p*ro*ductu*m* de t*ri*plicatiſ.
Vt q*ue* p*ri*uſ dixi nu*mer*uſ qui p*ro*ue*n*it in*de*.
A ſupra po*ſ*itiſ haſ reſpiciendo t*ra*hat*ur*.
Hi*nc* cubice iu*n*cta*m* ſup*pri*mo m*u*ltiplicato
Reſp*e*ctu q*ue* ſui remouebi*s* de remane*n*ti.

28 The scribe has corrected *adiemetur* to *adimeter*.

1 t*ibi*] sibi *H*, *Σ*, *Θ* **2** ſi] *om.* *Θ* **2** quid] quicquid *Θ* **3** pot*er*it] poteris *H*, *Σ* **3** minue*n*do] inveniendo *H* **4** Hi*nc*] hunc *H* **4** t*ri*plica*n*ſ] triplans *H*, *Σ*, *Θ* **4** ſaltu*m*] saltem *H* **4** illa] illo *H* **5** Q*ue*] Quod *H* **5** figura*m*] figura *H* **6** Illi] Sibi *H* **6** p*re*poneſ] propones *H*, *Σ* **6** ſociet*ur*] asocietur *H*, *Σ* **7** Vt] Et *H* **9** p*ro*ue*n*it] pervenit *H* **10** pone] tolle *H*, *Σ*, *Θ* **13** poſitiſ] omnes *Σ* **14** dele*n*ſ] dele *H*, demes *Θ* **14** t*ri*pleſ] triplans *Θ* **15** t*er*naſ] terna *H*, *Θ* **15** ret*ra*haſ] retrahens *H*, *Σ*, *Θ* **16** p*re*pone] propone *H*, *Σ* **18** ſub t*ri*pliciſ] sub triplis *H*, *Σ*; cum subtriplis *Θ* **18** subiu*n*ctiſ] sibi junctis *H*, *Σ* **19** de t*ri*plicatiſ] multiplicabis *Θ* **20** p*ro*ue*n*it] prevenit *H* **22** Hi*nc*] Huic *H*, *Σ* **22** iu*n*cta*m*] junctum *H*, *Θ*; ductum *Σ* **22** ſup*pri*mo] sub primo *Σ*

into itself cubically, eliminates that which you have placed above. And if what was **f. 77r** adjacent on the left side, if not entirely, it should be diminished as much as possible. After this, draw back three steps, making under these what continues with a digit by eliminating three. You should place the figure under that third with which it is associated, so the triple is multiplied by itself with the subtriple. After this you should multiply the product by it alone. Thereafter, place the total number which thence resulted in the position above in consideration of having been tripled, having been added at first to the cube which has been multiplied. In consideration of that, the number which will proceed from multiplying the cube will be taken away from that placed above. Then erasing the triple itself, in making a step, always by three, drawing back[21] another triple from the tripled, place a another figure here, which is multiplied by three in the prior manner. At first having joined the subtriple, thereafter by itself, multiply the product into the number with that which was tripled. As I said earlier, the number which results thence from those suppositions above should be positioned adjoining with respect to this[22] cube having been multiplied first. And with respect to itself, you should remove from the remaining

21 Reading *retrahens*, as in *H*, Σ, and Θ, for *retrahas*.
22 Reading *huic*, as in *H* and Σ, for *hinc*.

f. 77v *Et* ſic p*ro*cedaſ ret*ra*hendo t*ri*plando figura*m*.
Additiendo noua*m* donec tot*um* p*er*agatur.
ſubt*ri*pla ſub p*ro*p*r*iiſ ſ*er*uare dec*et* t*ri*plicatiſ.
Si nil in fine reman*et* nu*meru*ſ dat*us* an*te*.
Et cubic*us* cubicam radice*m* ſub t*ri*pla p*re*bent.
Cum digito iu*n*cta q*ue*m ſupp*rim*o poſuiſti.
H*ec* cubice ducta nu*merum* redd*et* t*ibi* p*rimum*.
Si q*ui*d erat remane*n*ſ no*n* e*ſt* cubic*us* ſ*ed* habe*tur*.
Maior ſub p*ri*mo qui ſtat radix cubicati.
S*er*uari q*ue* dec*et* q*ui*cq*ui*d radice rema*n*ſit.
Ext*ra*cta. nu*mer*o dec*et* h*ic* addi cubicato
Quo f*act*o nu*meru*ſ[29] deb*et* reddi t*ibi* p*rimus*.
Nam debeſ p*er* ſe radice*m* m*u*ltiplicare.
Ex hi*nc* i*n* nu*merum* duceſ qui p*ro*ue*n*it inde.
Sub p*ri*mo cubic*us* maior ſic in ue*n*ie*tur*.
Illi iu*n*gat*ur* remane*n*ſ et p*rimus* habe*tur*.
Si p*er* t*ri*plat*um* nu*merum* neq*ue*aſ op*er*ari.
Cyfra*m* p*re*poneſ. nil u*er*o p*er* ha*n*c op*er*are.
S*ed* ret*ra*he*n*ſ illa*m* cu*m* ſaltu dei*nde* t*ri*plata*m*
P*re*poneſ illi digit*um* ſub lege p*ri*ori.
Cu*mque* cyfram ret*ra*heſ ſalie*n*do n*on* t*ri*plicab*iſ*.
Na*mque* nic*hil* cyfre t*ri*plicat*i*o dicit*ur* eſſe.
At cu*m* tu cyfra*m* p*er*traxeriſ au*t* t*ri*plicata*m*.

29 *H* has *nnumerus* here, which is clearly a typographical error.

1 figura*m*] figuras *H* **2** Additiendo] Et proponendo *H*, *Σ*; preponendo *Θ* **2** noua*m*] nonam *H*, *Σ* **5** Et] Est *H*, *Σ*, *Θ* **5** p*re*bent] probent *H* **6** iu*n*cta] juncto *H*, *Σ*; sumpta *Θ* **6** ſupp*rim*o] sub primo *H* **7** H*ec*] huic *H* **7** redd*et*] reddant *H*, *Σ*; reddunt *Θ* **9** cubicati] cubicam *Σ* **10** q*ue*] *om.* *H*, *Σ*, *Θ* **10** dec*et*] debet *H*, *Σ*, *Θ* **11** Ext*ra*cta] Extracto *H*, *Σ*, *Θ* **11** h*ic*] hoc *H*, hec *Σ* **14** qui] quod *H* **14** p*ro*ue*n*it] pervenit *H* **18** p*re*poneſ] propones *H*, *Σ* **18** op*er*are] operari *H* **19** ret*ra*he*n*ſ] retrahaſ *Θ* **19** t*ri*plata*m*] triplata *Σ* **20** P*re*poneſ] Propones *H*, *Σ* **20** lege] more *Θ* **21** Cu*mque*] cum *Θ* **22** t*ri*plicat*i*o] triplacio *H*, *Σ*, *Θ* **23** At] Aut *H* **23** p*er*traxeriſ] protraxeris *Σ*, *Θ* **23** au*t*] autem *H* **23** t*ri*plicata*m*] triplicata *Σ*, triplatam *Θ*

and so you should proceed by drawing back the figure by threefold. By adding a f. 77v new figure as long as the whole shall not have been finished, it is correct to save the subtriples on their own. If nothing remains in the end, the number given originally is[23] a cube. They display the cube root under the triples, with the digit adjoining[24] what you had placed first. These, having been multiplied cubically, should return[25] to you the original number. It is not a cube if something remains, but it is found greater whereby the root cubed is less than the original. And it ought to be saved, whatever remains from the root, from the number extracted,[26] it ought to be added to the cube. With this having been done, the original number should be returned to you. For you should mutiply the root by itself, then multiply into that number, which thus produces the greater cube as will be found less than the original. The remainder should be added to this and the original is found. If you are unable to work with a number triple, you should put a cipher, certainly you can produce nothing by this. But, drawing back that threefold with the next step, you should place the digit in that as in the previous manner. Whenever you draw back a cipher in making a step, do not triple, for in fact nothing is said to be by the tripling of a cipher. But when you extract a cipher or a triple,

23 Reading *est*, as in *H*, Σ, and Θ, for *et*.
24 Reading *juncto*, as in *H* and Σ, for *iuncta*.
25 Reading *reddant*, as in *H* and Σ, for *reddet*.
26 Reading *extracto*, as in *H*, Σ, and Θ, for *extracta*.

f. 78r Ha*n*c cu*m* ſubt*ri*plo ſe*mper* ſeruare meme*n*to.
Si det *compo*ſitu*m* digiti t*ri*placio d*ebet*.
Illiuſ inſc*ri*bi digit*us* ſalie*n*do ſub illa.
Deleto digito que te*r*na dicit*ur* eſſe.
Iu*n*git*ur* articul*us*. cu*m* t*ri*plata p*re*eu*n*ti.
S*ed* facit h*unc* ſc*ri*bi p*er* ſe t*ri*placio p*ri*ma.
Que ſi det digitu*m* p*er* ſe ſc*ri*bi facit illu*m*.
Co*n*ſu*m*pto nu*mer*o ſi ſole ſi*n*t t*ibi* cyfre.
T*ri*plate p*re*pone cyfra*m* ſaltu*m* faciendo.
Cu*m* q*ue* cifra ret*ra*he t*ri*pla*m* ſaliendo cyfra*m* q*ue*.
P*re*ponenſ cyfre ſic p*ro*cedenſ op*er*are.
Si treſ ſiu*e* due ſerieſ ſi*n*t pone ſub una.
A dextriſ digitu*m* ſerua*n*do p*ri*uſ docume*n*tu*m*.[30]
Se*quitur* deprogreſſione
Si ſit co*n*tinua p*ro*greſſio te*rmin*uſ impar.
P*er* mai*us* mediu*m* totale*m* m*u*ltiplicato
Si par p*er* mediu*m* t*unc* m*u*ltiplicato ſeque*n*te*m*.
S*ed* ſi co*n*tinua no*n* ſit p*ro*greſſio finiſ.
Impar t*unc* mai*us* mediu*m* ſe m*u*ltiplicab*iſ*.
Si par p*er* mediu*m* ſibi m*u*ltiplicato ſeq*ue*nte*m*.[31]
Mvltiplica*n*s digitu*m* digito maior ſpecu*la*ri.
Qu*an*to ſtet deniſ altu*m* q*ue* ſuo duplicare.
Fer tocie*n*ſ c*re*ſce*n*ſ q*ue* ſere*n*ſ d*e* m*u*ltiplicare.

---

30 *H* ends here **31** Σ and Θ end here

---

1 Ha*n*c] Huic *H* **3** inſc*ri*bi] scribi *Σ*, Θ **3** ſub] super *H* **3** illa] ipsam *H*, *Σ* **4** Deleto digito que te*r*na dicit*ur* eſſe] Quæ manet a digito deleto, terna figura *H* **5** p*re*eu*n*ti] pereunti *H*, *Σ* **8** ſi*n*t] fuit *H*, *Σ* **8** t*ibi*] ibi Θ **9** T*ri*plate] triplato *Σ* **9** p*re*pone] propone *H*, *Σ* **10** cifra] cifram *H*, *Σ*; cifras Θ **10** t*ri*pla*m*] triplatam *H* **10** ſaliendo cyfra*m* q*ue*] scribendo figuram *H*, *Σ* **11** P*re*ponenſ] Propones *H*, Preponas *Σ*, Θ **12** ſiu*e*] vel *H*, *Σ* **12** due] duo *H*, *Σ* **12** ſerieſ] serie in *Σ* **12** una] yma *Σ*, ima Θ **15** impar] nuper *Σ* **19** ſe] si *Σ* **19** m*u*ltiplicab*iſ*] multiplicabit Θ, *B* **20** m*u*ltiplicato] multiplicabo Θ **20** ſeq*ue*nte*m*] propinquum *Σ*, Θ, *B* **21** ſpecu*la*ri] speculare *B*, *E* **22** altu*m* q*ue*] aliumque *B*, *E* **22** ſuo] sue *E* **22** duplicare] decuplare *B*, *E* **23** Fer] fert *E* **23** q*ue*] quam *B* **23** ſere*n*ſ] sciens *B*, scias *E* **23** d*e*] se *B*

always remember to save this with the subtriple. If it is a compound, it should be written of it by tripling of the digit, the digit having been stepped under it.[27] Erase after the digit, what thrice is said to be. The article is joined to the preceding triple, but it is to be written here by itself tripling the first. If that produces a digit, it is to be written by itself. When the number has been used up, if there should be only triple ciphers for you, place a cipher when making the step, at which time draw back to the third cipher,[28] that cipher having been stepped over. Thus proceeding to work, you shall place[29] ciphers. If there are three or two rows, place beneath the lowest[30] from the right the digit having been saved in the earlier instructions. **f. 78r**

Next about a progression

If the final term of a continuous progression is odd, multiply half of the total number of terms by the greater. If even, multiply the next by half of the total. But if the progression is not continuous, if the last term is odd, then you will multiply the greater middle by itself. If even, multiply by half of itself the next.

Multiplying a digit by a digit, observe[31] by how much the greater one remains from ten, and multiply the other one[32] by ten.[33] Take away as many times what arises and was brought forth from what was multiplied,

---

27 Reading *ipsam*, as in *H* and Σ, for *illa*.

28 Reading *cyfram*, as in *H* and Σ, for *cyfra*.

29 Reading *preponas*, as in Σ and Θ, for *preponenſ*.

30 Reading *ima*, as in Σ and Θ, for *una*.

31 Reading *speculare*, as in *B* and *E*, for *ſpeculari*.

32 Reading *alium*, as in *B* and *E*, for *altum*.

33 Reading *decuplare*, as in *B* and *E*, for *duplicare*.

f. 78v Nam p*ro*creſce*n*ti reſta*n*ſ debeſ reputare.
Si cupiſ artic*u*l*u*m pare m*u*tiplic*a*re meme*n*to.
Illo*rum* digitoſ quoſ p*er* ſe m*u*ltiplicaſti.
Hi*n*c q*u*od eru*n*t monadeſ tot centenoſ numerato.
Artic*u*lum digito te m*u*ltiplica*n*te notato.
Artic*u*li digi*tum* reliq*uo* q*ua*m m*u*ltiplicaſti.
P*ro* q*ua*muis monade creſcente q*u*od reputari.
In mixt*um* digitu*m* ſi q*uando* ducer*e* queriſ.
Illu*m* diſſolue*n*ſ digito*rum* docma ſeq*u*eriſ.
Articuli digitiq*ue* ſequi p*o*st iuſſa iuberiſ[32].
Artic*u*lum duceſ i*n* mixt*um* ſic op*er*atur.[33]
Mixt*um* ſe c*er*ne*n*ſ ia*m* d*ic*ta lege fruat*ur*.
Que fit in articuliſ hi*n*c nor*me* iuſſa ſeq*ue*riſ.
Int*er* ſe mixti. *cum* multiplicet*ur*. eo*rum*.
Duca*n*t*ur* digiti ſeu norma iub*et* digito*rum*.
Articuli p*o*st h*ec* uelut edoc*et* artic*u*lo*rum*
P*o*st et in artic*u*l*u*m reliqui quiuiſ digito*rum*
H*ec* ſimu*l* adde ſcieſ p*ro*duct*um* *com*poſito*rum*.
Et no*n* pulu*e*rea ſic duct*io* fit nu*mer*orum.
Explic*it* algoriſmus.

---

32 *B* has the misspelling *uiberis*. **33** *E* omits lines 11 through 13.

---

2 pare] *om. B* **2** meme*n*to] nominabis *B*, notabis *E* **3** quoſ] quoque *B* **4** Hi*n*c] hunc *E* **4** tot] tociens *B* **4** centenoſ] cetera *B* **6** Artic*u*li] Articulus *E* **6** m*u*ltiplicaſti] multiplicato *B*, multiplicaui *E* **7** q*ua*muis] quavis *B* **7** q*u*od] decem *B*, *E* **7** reputari] reputato *B*, reputate *E* **8** In] *om. B* **8** ſi] uero *E* **9** docma] dogma *B* **10** ſequi] sequens *B* **11** duceſ] ducens *B* **13** Que] Quod *B* **13** articuliſ] articuli *B* **13** nor*me*] ratione *B* **13** ſeq*ue*riſ] sequetur *B* **14** m*u*ltiplicet*ur*] multiplicantur **15** Duca*n*t*ur*] Mutantur *B* **15** ſeu] ceu *B* **18** ſcieſ] sciens *B*

for you should calculate the remaining from what has grown fully. f. 78v

If you wish to multiply an article by a similar number, take note of their digits which you will multiply[34] by themselves. Then there will be as many hundreds as what had been counted units. When you are multiplying an article by a digit, note the digit of the article which you will multiply according to whatever[35] multiplier unit is to be calculated. If ever you ask to multiply a digit into a mixed number, follow the teaching separating that into the digits of the article and the digit. Afterwards, you are directed to follow the instructions. You will multiply the article in the mixed number in such a way: it is performed separating the mixed number from itself. Now by the stated principle, it is fruituful that what is in the article[36] from here you will follow the standard instructions. When mixed numbers are to be multiplied[37] with one another, the digits should be multiplied in the same way[38] by the standard order of the digits, after these, the articles just as he teaches of articles. And after, whatever remains of the digits into the article. At the same time, add these to the product of the parts you know, and so the multiplication of the numbers is not covered with dust.

Here ends the alogorismus.

34 Reading *multiplicabis*, as in *B* and *E*, for *multiplicaſti.*

35 Reading *quavis*, as in *B*, for *quamvis*.

36 Reading *articuli*, as in *B*, for *articulis*.

37 Reading *mulitplicantur*, as in *B*, for *multiplicetur*.

38 Reading *ceu*, as in *B*, for *ſeu*.

7 Concerning the Geometry of a Circle: Transcript and Translation

7.1 The Transcription and Critical Apparatus

The following transcript is from GKS 1812 4to f. 13r. The text appears as well in AM 415 4to f. 9v, AM 685 d 4to ff. 29r–30r, and AM 736 III 4to ff. 2r–2v. Except for the version in AM 415, the exposition is accompanied by a diagram illustrating the relationships between the various geometric figures described in the text. The diagram in GKS 1812 is shown in figure 1.1 and is reproduced here in figure 7.1. The diagram in AM 685 d is similar, except that it omits the label "xxxiij" for the diameter of the outer circle. The diagram in AM 736 III is missing the numerical labels, has the square rotated 45 degrees, and has a man (apparently a monk) inscribed inside the triangle. The diagram in GKS 1812 has some additional markings, not reproduced here, in the lower left-hand side of the diagram which are likely from a pen test.[1]

Previous transcripts of the text may be found in *Alfræði Íslenzk*,[2] based primarily on AM 415 and AM 685 d, and in *Scriptores rerum danicorum medii ævi*,[3] based on AM 415. The latter also contains a Latin translation. Gustaf Eneström published a French translation in 1898.[4]

The scribe who copied this section of GKS 1812 is not the same as the scribe who copied the *Algorismus*. However, the two scribes were contemporaries who collaborated on some of their work, including jointly writing a different section of the manuscript. Whereas the scribe of the *Algorismus* was likely Norwegian or at least trained in Norway, this scribe's script indicates that he was Icelandic. Whether the collaboration between the two scribes took place in Iceland or Norway is impossible to infer.[5]

The general guidelines for the transcription are the same as those described above in Section 5.1 for the transcript of the *Algorismus*. There is one dittography in the text, denoted as ⊣ text ⊢. The critical apparatus does not reflect differences in the use of accents, the letters 'u' and 'v,' the letters 'i' and 'j,' punctuation, capitalization, or abbreviations. Where two or more manuscripts essentially agree, but differ from GKS 1812, the critical apparatus notes the reading from the oldest manuscript. The

1 See the first footnote in: Beckman and Kålund, *Alfræði Íslenzk*, 231.

2 Beckman and Kålund, *Alfræði Íslenzk*, 231–32.

3 Langebek, *Scriptores rerum danicorum medii ævi*, 192.

4 Eneström, "Sur quelques propositions de planimétrie," 19.

5 Haraldur Bernharðsson, "Scribes and Scribal Practice," 113.

https://doi.org/10.1515/9781501522376-007

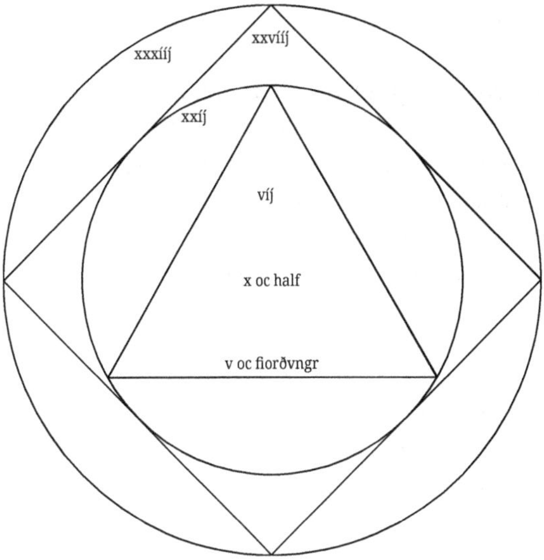

Fig. 7.1: Diagram as it appears in GKS 1812, f. 13r

line breaks are those of GKS 1812, with the exception that words are not broken at the end of lines. Given the brevity of the text, and the fact that it has not been studied as much as the *Algorismus*, the critical notes here are more detailed than those in the preceding texts.

In the critical apparatus, *Λ* is AM 415, *B* is AM 685 d, *Γ* is GKS 1812, and *Δ* is AM 736 III.

7.2 The Manuscripts: AM 415 4to (Arnamagnæan 415 quarto)

The manuscripts other than AM 415 4to were discussed earlier in Section 5.4. The manuscript AM 415 4to consists of twelve folios dating from the early part of the 14th century. The manuscript is now held at the Árni Magnússon Institute in Reykjavík. Only narrow strips of the first two folios survive. Along with the text on the geometry of a circle, there are lists of the names of kings of Norway, Denmark, and Sweden, as well as popes. There is a diagram of the planets and a text on the stars, along with the *Stjǫrnu-Odda tal*. The last folio has genealogies of kings from Noah and Odin and some history of the church of Hamburg.

More details on the manuscript and a bibliography may be found at handrit.is (https://handrit.is/manuscript/view/is/AM04-0415).

7.3 The Geometry of a Circle

f. 13r [V]m melingſ hv*er*ſ hringſ. *er* þrimlvtvm leng*ri* en breid h*anſ ok* ſiavndvngr af ⊣af⊢ en*ne* fiorðv breidd. þ[v]i at v*er*dr ef ʌ alna *er* breidd h*ring*ſinſ. þa *er* .ᒣᒣ. Ef þu g*er*ir f*er*ſkeytv iafna vm hring vtan. þa *er* um mẻling hen*nar* fiogr ſlik ſem breidd h*ring*ſínſ. þ*at* er .8. alnar ok ᒣθ. ef ʌ alna *er* breidd h*ring*ſinſ. Ferſkeyta hv*er* *er* þriðivngi breidari i ſkacc horn en vm þv*er*t. Nv gior*ir* þv h*ring* vtan vm horn f*er*ſkeytv. ok *er* vm meling ok breidd þeſſ h*ring*ſ. þridivngi leng*ri*. en hinſ h*ring*ſ *er* in*n*an v*ar* i f*er*ſkeytv. þa v*er*dr breidd h*an*ſ half elliftti alín. En vm mẻling h*an*ſ .∃5.[1] alna Ef þv g*er*ir þriſkeytv iafna i h*ring*ínu*m*. þa *er* af midri rỏnd vpp i hornít fiorðvngi ſcemra en breidd h*ring*ſinſ. Ef h*ring*r *er* .ʌ alna. þa *er* þriſkeyta fim alna. ok fiordvngr aln*ar*. En ſida þriſkeytv er half*rar* ſettv alnar.[2] ef breidd h*ring*ſ er half*rar* ſia*n*dv alnar.

1 Should be ∃∃.
2 *Δ* ends here.

1 [V]m melingſ] Ummæling *Λ, B, Δ* 1 hv*er*ſ hringſ] hringſ hvers *Λ*; hvers hringſ *B*; hvers hrings *Δ* 1 þrimlvtvm] þrim lutum *B*; þrimr lutum *Λ*; .3. *Δ* 1 leng*ri*] meíri *Δ* 1 breid] bréidd *Λ, B, Δ* 1 h*anſ ok*] ok *B* 1 ſiavndvngr] fiordungr *Δ*; ſia*n*dungr *Λ, B* 2 en*ne*] enni *Ω*; henne *B*; hínne *Δ* 2 fiorðv] fioðo *Λ*; ꝗ° *B* 2 breidd] *om. B* 2 þ[v]i at v*er*dr] þvi at sva verþr *Λ*; þviat sva verþr *B*; *om. Δ* 2 alna] alnar *Λ* 2 breidd h*ring*ſinſ] bréidd hrings *Λ, B*; hringur breidr *Δ* 2 ᒣᒣ] ᒣᒣ. alnar vm mæling hans *Λ*; 22 um mæling hanſ *B*. .22. alnar ummælíng hans *Δ* 3 um mẻling] ummæling *Δ*; um mǽling *Λ, B* 3 fiogr] fiogvr *Λ*; 4 *B, Δ* 4 ſlik] sklik *Λ* 4 h*ring*ſínſ] hringſ *Λ, B, Δ* 4 .8. alnar ok ᒣθ] .28. *B*; .28. alnar *Δ* 4 alna] *om. Λ*; alnar *B* 5 h*ring*ſinſ] hans *Λ*; hringſ *B*; hrings *Δ* 5 Ferſkeyta] fǽrſkeyta *Λ* 5 hv*er er*] hver er iofn er *Λ*; huer iofn er *B, Δ* 5 þriðivngi] þriðivgr *Λ* 5 breidari] breiðari *Λ*; breidare *B*; leíngrí *Δ* 5 ſkacc] ſkack *B, Δ*; ſkakk *Λ* 5 horn] hornn *Δ* 5 en] enn *Λ, Δ* 6 þv*er*t] þuer *Δ* 6 gior*ir*] gerir *Λ* 6 vtan] vttan *Λ* 6 horn f*er*ſkeytv] horn fer ſkeyto *Λ*; ferſkeytu horn *B* 6 er] *om. Δ* 6 vm meling] ummæling *Δ*; um mę̈ling *Λ*; um mæling *B* 7 þeſſ] þes *Λ* 7 h*ring*ſ] hrings *Δ* 7 þridivngi] þriðiungi *Λ* 7 leng*ri*] leingri *B*; meíre *Δ* 7 en] enn *Λ, Δ* 7 hinſ] hins *Δ*; þess *B* 7 h*ring*ſ] hrings *B*; *om. Δ* 7 f*er*ſkeytv] fer ſkeyto *Λ*; *om. Δ* 8 v*er*dr] verþr *Λ* 8 h*an*ſ] *om. Λ, Δ* 8 elliftti] ιι *Λ*; ellefta *B*; tolfta *Δ* 8 En] enn *Λ, Δ* 8 vm mẻling] ummæling *Δ*; vm mę̈ling *Λ*; um mæling *B* 8 h*an*ſ] hans *Λ*; *om. Δ* 8 .∃5.] ∃∃ *Λ, Δ*; ꝗᒣ *B* 9 g*er*ir] giorir *Δ* 9 iafna] iamn *Λ*; íafnna *Δ* 9 h*ring*ínu*m*] hring *Δ* 9 af] *om. B* 9 midri] miðri *Λ, B*; midre *Δ* 9 rỏnd] rond *Δ*; ra*n*d *Λ* 10 fiorðvngi] fíordvngi *B, Δ* 10 ſcemra] ſkemra *Λ*; skem *B* 10 en] enn *Λ, B, Δ* 10 breidd] bredd *Λ* 10 h*ring*ſinſ] hringſ *Λ*; *om. Δ* 11 þriſkeyta] ferſkeyta *B* 11 fim] ϥ *Λ, B, Δ* 11 fiordvngr] fiorþvngr *Λ* 11 En] enn *Λ, B, Δ* 11 ſida] ſiða *Λ* 12 ſettv] ნ *Λ* 12 h*ring*ſ] *om. B* 13 ſia*n*dv] ſia*n*do *Λ*; ნ *B*

7.3 The geometry of a circle

The circumference of any circle is three parts longer than its breadth plus a seventh part of a fourth breadth. So it is, if 7 ells is the breadth of the circle, then it is 22.[1] If you make a square about the circle from the outside, then its circumference is four such as the breadth of the circle. That is 8 ells and 20 if 7 ells is the breadth of the circle. Any square is by a third part broader than around through opposite angles. Now make a circle outside around the angles of the square, and around the perimeter, and the breadth of this circle is by a third part longer than of the circle which was inside the square. Then the breadth of it is half eleven[2] ells. And its circumference is 35[3] ells. If you make an equilateral triangle in the circle, then it is from the middle of an edge up to the angle a fourth shorter than the breadth of the circle. If the circle is 7 ells, then the triangle is five and one-fourth ells. And a side of the triangle is half six[4] ells, if the breadth of the circle is half seven[5] ells. f. 13r

1 AM 415 adds "ells around its circumference."
2 That is, 10.5.
3 Should be 33.
4 That is, 5.5.
5 That is, 6.5.

8 Practical Arithmetic and Geometry

The philosophy of mathematics has long played a central role in the study of philosophy as a whole. Unlike the philosophy of biology, the philosophy of physics, or many other similar subdisciplines of philosophy, the philosophy of mathematics has, since at least the time of Plato, been seen as key to addressing many questions at the heart of inquiries into the nature of human knowledge. This is so much the case that the philosopher Ian Hacking wrote a book to examine why this is so.[1] Hacking focused his answer on two notable aspects of mathematical thought: the central role of proof in determining mathematical truth, and the fruitfulness with which the results of such an abstract science may be applied to the workings of the natural world. In a sense, mathematics provides a bridge between a world of abstract objects, a world in which it is possible to discern properties which are necessarily true—that is, cannot be otherwise—and the sensible world of contingent beings. Philosophical answers to the question of how this bridge can exist, and what this means for the nature of the objects on either end of it, span the gamut of philosophical thought on how and what we can know. At one extreme, we have have the platonic view that what is really real is the world of forms on one end of the bridge, with the other side consisting of only the shadows of the real At the other extreme is the social constructivist view that only the contingent physical objects are real, with the other side of the bridge consisting of nothing more than the agreed upon objects of human thought. Here, we look at a single question which resides within the much larger question of the nature of mathematical thought: what is the place of algorithmic thinking, as exemplified in the *Algorismus*, within the realm of mathematical thought?

8.1 Practical and Speculative Mathematics

Historically, mathematics has always had two closely related, yet distinct, branches. One, often referred to as pure mathematics, studies mathematical objects and structures for their own sake. Although applications to the physical world might follow later, as, for example, the general theory of relativity required the prior development of non-euclidean geometries, pure mathematicians pursue their subject for its own sake. The second branch, applied mathematics, explicitly aims to develop mathematical knowledge for the purpose of utilizing it to provide better models for

1 Ian Hacking, *Why is There Philosophy of Mathematic at All?* (Cambridge: Cambridge University Press, 2014).

applications outside of the realm of mathematics itself. Applied mathematics is no less mathematical than pure mathematics. Proof is as important in applied mathematics as it is in pure. The difference lies in the extent to which the direction of research is motivated by applications outside of mathematics itself, the second of Hacking's notable aspects of mathematical thought.

In the Middle Ages, another distinction between branches of mathematics was drawn, similar to our modern classification of pure and applied mathematics, yet along different lines. Medieval thinkers saw a distinction between the parts of mathematics—the speculative—which studied properties of mathematical objects and structures, and the relationships between them; and those parts—the practical—which explained how concrete computations might be performed to determine numeric values for certain properties and relationships. For example, it is one thing show that that the ratio of the circumferences of two circles is the same as the ratio of their diameters, as Euclid demonstrated, or even that the constant of proportionality between the circumference of a circle and its diameter is between $\frac{223}{71}$ and $\frac{22}{7}$, as Archimedes showed. However, it is another to simply use $\frac{22}{7}d$ as an acceptable approximation for the circumference of a circle of diameter d. The first results are speculative, the latter practical. Similarly, the fact that the height of an equilateral triangle with sides of length s is $\frac{\sqrt{3}}{2}s$ becomes practical only when one has a good approximation for $\sqrt{3}$. For example, as we saw earlier, one might follow Gerbert in computing the height as $\frac{6}{7}s$.

In discussing the divisions of arithmetic, the anonymous author of the 13th-century *Philosophica disciplina* first defines arithmetic and then divides it into the theoretical and the practical:

> Arismetica, que est prima pars, diffinitur sic : Arismetica est disciplina quantitatis numeralis secundum se.
> Diuiditur autem prima sui diuisione in theoricam et practicam. Theorica enim considerat numerum in se secundum suam essentiam, non prout est in materia, set secundum quod accidit ei quod sit par, uel superfluus, uel diminutus et huiusmodi. Practica autem considerat numerum in materia secundum quod accidit ei congregari, et disgregari, et similia; et docetur in Algorismo.[2]
>
> [Arithmetic, which is the first part, is defined as follows: Arithmetic is the study of numeric magnitude with respect to itself.
> Moreover, it is divided first into theoretical and practical divisions of itself. Namely, the theoretical considers number in itself with respect to its essence, not as it is in matter, but according to what belongs to it insofar as it is equal, or excessive, or deficient and of this sort. On the other

2 Claude Lafleur, ed., "Philosophica disciplina," in *Quatre introductions à la philosophie au XIIIe siècle* (Montreal: Institut d'Études Médiévales, 1988), 266.

> hand, practical considers number in matter according to what belongs to it to be collected, and to be divided, and the like; and it is taught in *algorismus*.]

That is, the theoretical, or speculative, side of arithmetic considers the properties of numbers and their relationships. The example property referred to here concerns whether or not a number is equal to, less than, or greater than the sum of its divisors. For example, the divisors of 28 are 1, 2, 4, 7, and 14, and so 28 is equal to the sum of its divisors. Such a number is said to be *perfect*. On the other hand, the sum of the divisors of 12 is greater than 12 (that is, $1 + 2 + 3 + 4 + 6 = 16$), and the sum of the divisors of 8 is less than 8 (that is, $1 + 2 + 4 = 7$). In the former case, we say that 12 is *excessive*, and in the latter we say that 8 is *deficient*. Perfect numbers have few, if any, practical uses, yet there are open questions about them that are still studied today.[3]

On the other hand, the arithmetic of the *Algorismus* is clearly of a very basic practical nature: the text explains how to add, subtract, multiply, divide, and perform other such operations with numbers. Most often, such calculations are performed in regard to numbers representing physical quantities, such as length, weight, or quantity, and so, in a practical sense, refer to numbers "in matter." The fact that the spread of the system of Indo-Arabic numeration is due largely to merchants, traders, and astronomers attests to the practical side of this calculus. This distinction between speculative and practical can even be seen in the description of even and odd numbers. As noted earlier, the classical definition of an even number is that a number is even if it is divisible into two equal parts. This definition is seen, for example, in the 14th-century *Arithmetica speculativa* of Thomas Bradwardine:

> Par numerus est qui in duo equalia nullo medio intercidente diuiditur. Impar qui sine intercidente vnitate in duo equalia nusquam dvidi potest et hec est diffinitio vulgaris.[4]
>
> [An even number is what is divisible into two equals with nothing intermediate interceding. An odd number is what is never able to be divided into two equals without an interceding unity, and this is the common definition.]

But we read something very different in the *Algorismus*:

> hueria tolu er þu ritar þa er hon íofnn ef tígum gengnir. edir jafnn fíngir er umfram En oll tala er oiofnn. ef oiafnn fingir er umfram Jafnnir fingir eru fiorír. ᒣ.ꝶ.ɕ.8. En oiafnnir. aðrir fiorir .Э.Ч.Λ.9. En eiɴ er huarki þuí at han er eigi tala heldir uphaf allrar tolu.[5]

3 For examples, it is not known if there exists an odd perfect number or if there are an infinite number of perfect numbers.

4 Thomas Bradwardine, *Arithmetica speculativa* (Paris: Guy Marchand, 1496), 1.

5 GKS 1812 4to, f. 13v, ll. 19–22.

> [A number is even if it is made from groups of ten or if an even digit is in the front. A number is odd if the digit in front is odd. There are four even digits: 2, 4, 6, and 8. And the other four are odd digits: 3, 5, 7, and 9. But one is neither because it is not a number but rather the origin of all number.]

Hence a speculative work on arithmetic tells us what an even number is in terms of its relationship with other numbers: namely, that it can be divided into two equal parts. This is in essence what an even number is, as reflected in the name given to such numbers in Latin, Icelandic, and English. Moreover, this definition does not depend on any method of representing the number, whether written in some representational system or depicted using markers or beads. A practical work tells us directly how to recognize an even number in some representational system, in this case the one given by the Indo-Arabic system: if the first digit on the right of the representation is even—that is, 2, 4, 6, or 8, or a cipher.

8.2 Knuth and Algorithmics

In the 5th century, Proclus summarized the ancient Greek understanding of what constituted mathematics:

> The Pythagoreans considered all mathematical science to be divided into four parts: one half they marked off as concerned with quantity (*ποσόν*), the other half with magnitude (*πηλίκον*); and each of these they posited as twofold. A quantity can be considered in regard to its character by itself or in its relation to another quantity, magnitudes as either stationary or in motion. Arithmetic, then, studies quantities as such, music the relations between quantities, geometry magnitude at rest, spherics magnitude inherently moving.[6]

These four areas of study, arithmetic, geometry, music, and spherics (that is, astronomy), would become the quadrivium, forming, along with the trivium of grammar, logic, and rhetoric, the core of medieval education.

Our notion of what topics are included within the mathematical sciences has changed over time. Music, as such, is no longer considered a part of mathematics (although the general study of relations between quantities is very much still at the core of mathematics) and much of the study of magnitudes in motion would now fall under the umbrella of physics. In part, these changes arise as distinctions are made between disciplines which pursue mathematics in itself and those which use mathematics as a tool to model phenomena outside of itself. Hence, for exam-

6 Proclus, *A Commentary on the the First Book of Euclid's Elements*, trans. Glenn R. Morrow (Princeton, New Jersey: Princeton Universy Press, 1970), 29–30.

ple, arithmetic[7] studies the structure of the integers as they relate to one another, whereas music studies relationships between integers as they relate to producing certain types of sounds.

The place of algorithms, including their creation, implementation, and the study of their efficiency, within the disciplines of the mathematical sciences has undergone reconsideration as well. As one notable example of this direction of thought, in 1979 an international group of prominent mathematicians and computer scientists gathered in Khiva in Uzbekistan, the modern site of al-Khwārizmī's home city, Khwarezm, for a symposium dedicated to a discussion of the role of algorithms in mathematics and computer science. Donald Knuth, one of the organizers of this pilgrimage to Khiva, addressed the question of how algorithms fit into the study of mathematics. For his discipline of computer science, Knuth thought that a better name would be "algorithmics," thus emphasizing that the focus of inquiry is the algorithm, not the machine upon which the algorithm is implemented. For this presentation, he began with a statement of his desire to start a philosophical discussion: "My purpose in this paper is to stimulate discussion about a philosophical question that has been on my mind for a long time: What is the actual rôle of the notion of an algorithm in mathematical sciences?"[8] Instead of first defining mathematics and then testing to see whether or not the study of algorithms fits this definition, he considered whether or not those who are acknowledged to be mathematicians treat algorithms in the same way as those in his discipline of algorithmics.

8.2.1 What is Mathematics?

In response to the question "What is mathematics?," Knuth replies that "[m]athematics is what mathematicians do."[9] Although this may sound somewhat simplistic, as well as circular, Knuth is not alone in approaching the question in this manner. For example, Terence Tao, perhaps the most well known mathematician of this century, has said:

> My own personal definition of mathematics is more descriptive than prescriptive: mathematics is what mathematicians do, and mathematicians are the people who study mathematics. This sounds like a tautologically circular definition, but I think of it more as describing the

---

7 In the traditional sense, often called 'number theory' in modern mathematics.

8 Donald E. Knuth, "Algorithms in Modern Mathematics and Computer Science," in *Algorithms in Modern Mathematics and Computer Science*, ed. E. P. Ershov and D. E. Knuth (Berlin and Heidelberg: Springer, 1981), 82.

9 Knuth, "Algorithms in Modern Mathematics," 88.

> equations of motion of a stochastic dynamical system, in which breakthroughs in mathematics attract more attention by mathematicians, which in turn redefines the relative weighting of different fields of mathematics.[10]

Note, however, that neither Knuth nor Tao is trying to explicate that nature of mathematical objects or the way in which we reason about them. Rather, they are aiming at a practical matter of fact at the center of the discussion: Whether mathematical objects are objective or subjective, platonic ideals or social constructions, the discipline of mathematics is certainly a creation of academic society. That is, the general academic community considers certain intellectual activities, on the whole, to be, and certain others not to be, part of mathematics. No one in the community doubts that analyzing the distribution of prime numbers among the integers is part of mathematics, whereas all would agree that analyzing the relationship between Aeneas and Dido is not. Moreover, as noted above, a glance at the history of mathematics reveals how the areas accepted as part of mathematics have changed: many topics which were once part of mathematics, such as music and astronomy (or even astrology), are no longer included in the discipline, while modern abstract areas, such as category theory or model theory, have been added. Whether or not the objects and methods of mathematics are socially constructed, it is certainly true that what constitutes the academic discipline of mathematics is. Hence, in essence, Knuth replaces the intractable question of the definition of mathematics with a practical rule: we can assess what is mathematics by observing what it is mathematicians do. What mathematicians do—in particular, what good mathematicians do–has changed over time with the refinement of the notions which lie at its core and the winnowing away of those which do not. One insight from this approach is that we can identify those activities no longer considered to be part of mathematics. That is, we can a distinguish from the central core of mathematical investigation what is no longer considered to be a part of mathematics. In a sense, Knuth is saying that he may not be able to tell us what mathematics *is*, but, through an examination of what mathematicians do and, just as importantly, don't do, he can give us a sense of what it *is not*.

So Knuth's task now entails considering how practicing mathematicians work with algorithms. Examining a number of mathematical texts somewhat at random, he comes to an unsurprising conclusion: Much of mathematics, from calculus to analysis to algebra to topology, involves outlining procedures or processes which lead to a solution for some problem, the construction of some desired object, or, perhaps, to the verification that a certain object satisfies a certain definition. What

10 Terence Tao (@tao), "My own personal definition of mathematics," December 5, 2024, accessed December 10, 2025, https://mathstodon.xyz/@tao/113601397729489852.

might seem at first surprising, but on reflection is not, is his realization that the most algorithmic of all the texts he considers is Errett Bishop's *Foundations of Constructive Analysis.*[11] In this text, Bishop attempted to reconstruct as much of modern abstract mathematical analysis as he could using proof techniques involving explicit constructions of the relevant objects. In his view, much of modern mathematics is idealistic, based on notions of proof and existence which provide no insight into how the mathematical objects involved could be produced. He did not regard such mathematics to be of no worth, just in need of more rigorous formulation and proof. For the constructivist mathematician, the realm of mathematics extends beyond the positive integers only as far as the finite methods of finite beings can extend it. As Bishop emphasizes, anything beyond these finite constructions can have real meaning only for some type of infinite being:

> In the words of Kronecker, the positive integers were created by God. Kronecker would have expressed it even better if he had said that the positive integers were created by God for the benefit of man (and other finite beings). Mathematics belongs to man, not to God. We are not interested in properties of the positive integers that have no descriptive meaning for finite man. When a man proves a positive integer to exist, he should show how to find it. If God has mathematics of his own that needs to be done, let him do it himself.[12]

Although impressive, and of great philosophical interest, Bishop's reconstruction of mathematical analysis was seen by most mathematicians to be impractical. In particular, the constructions could be far more complex than the proofs of the so-called idealistic mathematics and the results of important theorems were often of much restricted scope.

Nevertheless, with its philosophical commitment to specifying step-by-step procedures for all aspects of mathematics, constructive mathematics was precisely where Knuth could most clearly see what separates computer science, the science of algorithms, from mathematics. First, a constructive proof must show only that the process terminates, in a finite number of steps, with the desired object and has no need to be concerned with the complexity or efficiency of the computation required to get there. Second, unlike the algorithms designed for implementation on a computer, the constructions of constructive mathematics are not truly dynamic. That is, new objects are created as the constructive process proceeds, but there is no sense, or need, of a new object replacing a previous one. In the algorithmics of Knuth, the concept of assignment, the replacement of a previously computed object by a new one, is essential. The constructive mathematician may have an indefinite

11 Errett Bishop, *Foundations of Constructive Analysis* (New York: McGraw-Hill, 1967).
12 Bishop, *Foundations of Constructive Analysis*, 2.

space to work in, but the computer scientist must always work within inflexible limits.

8.2.2 What is an Algorithm?

As V. A. Uspensky and A. L. Semenov pointed out at the conference in Khiva, perhaps "[t]he most important discovery in the science of algorithms was undoubtedly the discovery of the general notion of algorithm itself as a new and separate entity."[13] That is not to say that algorithms did not exist before they were identified as objects worthy of study in themselves, nor that that this recognition did not occur until the formalizations of notions of computational models in the 1930s. Nevertheless, it is clear that the recognition of an algorithm as a specific type of entity began to take hold at some point in the early part of the 20th century. For example, in 1912 Émile Borel defined a "calculable number" α as one for which it is possible, for any given positive integer n, to obtain a rational number which differs by no more than $\frac{1}{n}$ from α. He leaves it for a footnote to describe what such a process would entail:

> Je laisse intentionnellement de côté la plus ou moins grande longueur pratique des opérations; l'essentiel est que chacune de ces opérations soit exécutable en temps fini, par une méthode sûre et sans ambiguité.[14]
>
> [I am intentionally setting aside the practical length of the operations, whether greater or smaller; the essential point is that each of the operations should be executable in a finite time by a method definite and without ambiguity.]

This is not far from the definition Knuth uses in his paper, adding only that, for a general description, one must include the structure of the underlying data: "I tend to think of algorithms as encompassing the whole range of concepts dealing with well-defined processes, including the structure of data that is being acted upon as well as the structure of the sequence of operations being performed."[15] These informal characterizations of an algorithm will suffice for our purposes.

---

13 V. A. Uspensky and A. L. Semenov, "What are the Gains of the Theory of Algorithms: Basic Developments Connected with the Concept of Algorithm and with its Appplications in Mathematics," in *Algorithms in Modern Mathematics and Computer Science*, ed. E. P. Ershov and D. E. Knuth (Berlin and Heidelberg: Springer, 1981), 104.

14 Émile Borel, "Le calcul des intégrales définies," *Journal de Mathématiques Pures et Appliqueées 6ᵉ série* 8 (1912): 162.

15 Knuth, "Algorithms in Modern Mathematics," 82.

8.3 Algorithms and the *Algorismus*

The procedures for numeration, addition, subtraction, duplation, mediation, multiplication, division, and extraction of roots described in the various texts of the Indian calculus satisfy Borel's description of an algorithm. Each step of each procedure is specified without ambiguity, with all cases being considered, and requires only a finite number of operations to complete. Moreover, each method specifies the structure of the input data and how that input is acted upon. Indeed, the procedure for numeration provides the basis for how the input data are to be structured for each of the procedures which follow. Moreover, the methods satisfy Knuth's notion of assignment, with each step of a given procedure including instructions on how the parts of the original input data are to be replaced with the updated computations.

In other words, the procedures of the *Algorismus*, and the other texts of the Indian calculus, are described in a manner that is amenable to implentation on a machine. Of course, this is not surprising, as the methods are intended for implementation on physical objects such as dust boards or wax tablets, and are clearly related to the analogous procedures which might be performed manually on an abacus or counting board.

Some authors have contended that not all the procedures of the *Algorismus* are algorithmic. For example, Jens Høyrup, in a discussion of the procedure for finding square roots found in the work *Miftāh al-hisāb* by Jamshīd al-Kāshī, contends that the second step in the procedure is not strictly algorithmic. This step requires that one find the largest digit N satisfying a certain condition which depends upon the first step in the procedure. Høyrup contends that "[u]nspecified experimentation is hardly to be considered algorithmic in the strict sense."[16] Setting aside why Høyrup does not note that the same comment would apply to the first step of the procedure (where one must find the largest digit N for which N^2 is less than or equal to the first one or two digits of the given number), this step does not require unsepecified experimentation. Rather, it may be implemented with a simple loop to step through the digits until the desired N is found.

Moreover, there is another algorithm in the *Algorismus* which is not frequently noted: the determination of whether a given integer is even or odd. As noted above, the *Algorismus* tells us that whether or not a given number is even or odd depends entirely on whether or not the first digit, starting from the right, of the number is

---

16 Jens Høyrup, "When is the Algorithm Concept Pertinent – and When not? Thoughts About Algorithms and Paradigmatic Examples, and About Algorithmic and Non-Algorithmic Mathematical Cultures," *AIMS Mathematics* 3, no. 1 (2018): 220.

even or odd. This is, in fact, a simple algorithmic procedure for determining the parity of an integer, and one specific to having an Indo-Arabic representation of the number. In a modern implementation, this algorithm becomes even simpler, reducing it to simply looking at the lowest bit of the binary representation of the number and noting whether it is a 0 or a 1.

8.4 What Mathematics Isn't

To date, there has not been a fully satisfactory philosophical theory to explain and justify the nature of the objects and proofs of mathematics. In the early part of the 20th century, the debates centered on several approaches to the question, including attempts to formalize mathematics into nothing more than than the manipulation of certain symbols according to specified rules; attempts to reduce mathematics to a part of logic; and attempts to explicate a platonic view of mathematics as a science of a real, existent, immaterial world of ideal forms. Over the succeeding years, the debates expanded to include versions of constructivism, as seen, for example, in Bishop's work in analysis; structuralism, the notion that existence of mathematical objects lies in their position in various types of structures—that is, in their relationship to other parts of some structure; naturalism, the idea that mathematical objects represent phenomena that emerge from the properties of the natural world; social constructivism, the belief that mathematical objects and methods are the result of social conventions and agreements; and developments in realistic interpretations, including those based on platonism. That none of these has been entirely successful should not be surprising. After all, an answer to the central questions of mathematical philosophy would resolve many, if not most, of the fundamental questions of philosophy with regard to the nature of knowledge and the universe. Moreover, as we have seen above, there is not even general agreement as to which parts of human knowledge should falll under the heading of mathematics.

In the end, we can do no better than say, with Benjamin Peirce, that "[m]athematics is the science which draws necessary conclusions."[17] It follows, as noted by his son, Charles Peirce, that "[m]athematics is the study of what is true of hypothetical states of things."[18] For inferences about contingent things can never be necessary; it is only when when we are reasoning about things abstracted from all contingencies that our conclusions become necessary. Hence, for example, when

17 Benjamin Peirce, *Linear Associative Algebra* (New York: D. Van Nostrand, 1882), 1.

18 Charles S. Peirce, "The Simplest Mathematics," in *Philosophy of Mathematics: Selected Writings*, ed. Matthew E. Moore (Bloomington and Indianapolis, IN: Indiana University Press, 2010), 27.

creating an algorithm for a particular computation, and implementing that algorithm on a particular machine, one is acting as a computer scientist, or, as Knuth might prefer, an algorithmist. However, when studying an algorithm as an object in itself, perhaps to compare its efficiency relative to other algorithms of a particular type, one is then acting as a mathematician. What distinguishes those areas of study once (but no longer) considered by Proclus and others to be part of mathematics from those that still are considered part of mathematics (and those areas that have since been added) is the nature of the objects studied and the necessity of the conclusions drawn about them. So music, as a discipline, is no longer considered part of mathematics, but parts of the study of music, for example, the use of proportions in undestanding harmonies, are mathematical. Similary, astronomy is now considered to be a science separate from mathematics, yet every student of astronomy has need of mathematical tools.

Even if the objects of mathematical study are not socially constructed, the discipline of mathematics certainly is. Thus, over time, we have come to recognize that a discipline once considered under the umbrella of mathematics is, by its nature, not mathematical. The parts which consider abstract objects, such as numbers and their ratios, or the structure of space-time geometry, are subsumed into subdisciplines of mathematics, and the rest, the parts that work with the applications to the practical manifestations of these objects in the world, become disciplines unto themselves. Works such as the two texts from GKS 1812 treated here are found perhaps to belong more appropriately to the history, or maybe pre-history, of modern algorithmics. That does not lessen the importance of the texts, as the spread of the knowledge of how to to implement efficient algorithms for working with the Indo-Arabic system of numeration was valuable for all who had to keep accounts, engage in commerce, or perform other involved numerical calculations, whether they were monks, merchants, farmers, or scientists.

Yet to point to the *Algorismus* as evidence that Icelanders of the 13th and 14th centuries "kept up with the latest mathematical knowledge in the world," as some have contended, is to overstate the mathematical significance of the work.[19] One would need to look more toward the advances of Leonardo Pisano or the Oxford calculators to witness the state of mathematical knowledge of the period. Still, it is noteworthy that when Björn Gunnlaugsson wrote *Tölvísi* in the 19th century, the first attempt at a comprehensive mathematical textbook in Icelandic, covering topics from numerical computations through Fermat's little theorem and continued

19 Kristín Bjarnadóttir and Bjarni V. Halldórsson, "The Norse Treatise Algorismus," in *Actes du 10è, colloque maghrébin sur l'histoire des mathématiques arabes*, ed. Mahdi Adbeljaouad (Tunis: L'Association Tunisiennes des Sciences Mathématiques, 2010), 76.

fractions, he cited the *Algorismus* as a reference for the origin of the Indo-Arabic numbers.[20] This, in itself, reflects the place of respect this text held for Icelandic scholars even 500 years later. Nevertheless, a treatise on efficient techniques for working with numbers may be an important contribution to the advancement of practical mathematics, or algorithmics, without being at the same time at the forefront of the advancement of mathematical knowledge.

20 Björn Gunnlaugsson, *Tölvisi* (Reykjavik: Hið íslenzka bókmenntafélag, 1865), 4.

9 Mathematics and Numerology

> Sed omnia in mensura et numero et pondere disposuisti.
> [But you have ordered all things in measure, number, and weight.]
>
> — Wisdom 11:21[1]

> She knew that in the end you really cant know. You cant get hold of the world. You can only draw a picture. Whether it's a bull on the wall of a cave or a partial differential equation it's all the same thing.
>
> — Cormac McCarthy[2]

Looking to numbers, and mathematics more generally, to describe and explain the order of the natural world has a long history. From the beginnings of keeping accounts through enumerating quantities and measuring lengths, areas, and volumes; to the early discoveries of geometric and arithmetic procedures for computations and the important roles of proportions and ratios; to the modern applications of differential equations and differential geometry, natural scientists and mathematicians have sought to capture the structure of reality in an intelligible system of objects and relations.

What sets the Old Norse *Algorismus* apart from all other versions of the Indian calculus, and any manner of practical mathematics, is the inclusion of the neoplatonic identification of the four elements with the geometric progression between the first two cubes. We can do no more than speculate as to what prompted this inclusion. One can hardly imagine that it was due to some particular interest in platonic cosmology in 13th-century Iceland. It may be nothing more than that the source manuscript containing the *Carmen de algorismo* had a fragment of a commentary on the *Timaeus* immediately following. The discussion of cubes and cube roots in the original text may have been enough to bring this well-known neoplatonic exposition to the mind of the translator, or perhaps he simply saw it as a natural extension to the discussion of the geometric progression seen in other versions of the Indian calculus.

9.1 Using Mathematics to Model the World

Whatever the reason, we should not dismiss the passage as nothing more than ancient numerology. It is commonplace today that mathematics underlies our theories

---

1 *Biblia Sacra Iuxta Vulgatam Versionem*, 1016.

2 Cormac McCarthy, *The Passenger* (New York: Alfred A. Knopf, 2022), 279.

of the structure of the physical world. Some scientists and philosophers even go further, speculating that that structure may already be implicit within the properties of the integers themselves. For example, in a speech in 1939, Paul Dirac, one of the founders of the modern theory of quantum mechanics, said:

> There is thus a possibility that the ancient dream of philosophers to connect all Nature with the properties of whole numbers will some day be realized. To do so physics will have to develop a long way to establish the details of how the correspondence is to be made. One hint for this development seems pretty obvious, namely, the study of whole numbers in modern mathematics is inextricably bound up with the theory of functions of a complex variable, which theory we have already seen has a good chance of forming the basis of the physics of the future. The working out of this idea would lead to a connection between atomic theory and cosmology.[3]

Here Dirac is saying much more than just that mathematics underlies our understanding of nature: He is claiming that, in fact, properties of integers alone may suffice to explain it. The connections between the positive integers, more particularly, the prime numbers, and physical quantities, such as the energy levels of atomic nuclei, have become even more evident in recent years, with the discovery of deep connections between the two via the Riemann zeta function.[4]

Of course, the correspondence between properties of numbers and the natural world to which Dirac alludes is far deeper, mathematically, than the correspondence between the four elements and the geometric sequence described in the *Timaeus*. Yet, as Dirac argues, before the advent of modern theories of space-time and particle physics, the search for simple mathematical models of nature dominated the view of physicists.[5] Moreover, the philosophical underpinnings are similar: both posit a fundamental connection between relationships between integers and relationships between the objects of the study of physics. This is not to say that the natural world *is* mathematical. Rather, it is to say that the best, and perhaps only, way we have to grasp the world and make it intelligible is through mathematics. Calcidius, for example, encounters this distinction when he asks how it is that Plato could at one point in the *Timaeus* represent the four elements with geometric cubes and boxes and at another identify them with four of the platonic solids:

3 Dirac, "The Relation Between Mathematics and Physics," 129.

4 First postulated by Bernhard Riemann in 1859, the Riemann hypothesis specifies the location of complex numbers s for which $\zeta(s) = 0$, where ζ is the zeta function. The conjecture has important consequences for the distribution of prime numbers and remains the most outstanding unresolved question in mathematics.

5 Dirac thought the search for mathematical beauty had replaced the search for mathematical simplicity: Dirac, "The Relation Between Mathematics and Physics," 123.

> Dicet aliquis duo quidem solida corpora interuallo longiore distantia coniungi duabus interiectis medietatibus iuxta rationem continui competentis sufficienter probari sibi, sed inter ignem et terram, quae sunt solida corpora, nullam apparere similitudinem, quando iuxta ipsum Platonem ignis quidem forma et figura pyramoides esse dicatur, id est in modum pyramidis excresact, terra uero cubus sit, hae porro formae nullam ex se simiilitudinem mutuentur, quia non sint aequalibus angulis (quippe cubus omnis rectis est angulis, ergo, quia terra in cubica est figura, rectos angulos habeat necesse, at uero pyramidis anguli productiores sunt in acumine); ubi autem non sunt aequales anguli, ne latera quidem erunt pro competenti proptereaque similia sui non erunt. Quae distantia corpora si alia duo solida interiecta continuabunt, impedietur hoc pacto continuatio, siquidem inter duo non quaelibet sed quae similia sunt inuicem sibi solida corpora inseri debent.[6]

> [Someone will say that, although he is satisfied with the proof that two solid bodies separated by an extended interval are conjoined by two intermediates intercalated according to the ratio of a continuous proportion, nevertheless there is no apparent similitude between fire and earth, which are solid bodies. For according to Plato himself, fire is claimed to be pyramidical in form and shape, i.e., its extension is in the manner of a pyramid, whereas earth is a cube, and these shapes share no likeness with one another, since they do not consist of equal angles (for every cube consists of right angles, and so earth, being cubic in shape, necessarily has right angles, whereas the angles of a pyramid are more acute at the tip); and wherever the right angles are not equal neither will the sides be proportional, and hence they will not be similar to one another. It two other bodies are are to bind the discrete bodies by their intercalation, then on the current supposition the continuous bond will be impeded, since they should not be inserted between any two solid bodies whatsoever, but between ones which are similar to one another.[7]]

To resolve this apparent contradiction, Calcidius draws a distinction between what it is the two different geometrical descriptions are, in fact, describing:

> [M]emores nos esse debere eorum quibus Plato hanc ipsam difficultatem multo ante praeuidens eos qui ita sensuri sint ab errore reuocauerit. Dixit enim, si meminimus, similitudinem non solum in formis et figuris sed etiam in potentiis et qualitatibus.[8]

> [We must bear in mind the points by which Plato, who foresaw this difficulty well in advance, drew those inclined to take such a view away from their mistake. For he indicated, as we recall, that similitude is to be sought not only in forms and shapes but also in potencies and qualities.[9]]

Hence, whereas the platonic solids provide a geometric description of the four elements with regard to their appearances, cubes and boxes provide a way to describe their qualities.

---

6 Calcidius, *Timaeus: A Calcidio translatus*, XX, 71.
7 Calcidius, *On Plato's Timaeus*, 20, 149–51.
8 Calcidius, *Timaeus: A Calcidio translatus*, XXI, 71.
9 Calcidius, *On Plato's Timaeus*, 21, 151.

In a remark noted earlier in Section 3.1, Proclus sees the error that Calcidius is avoiding even more succinctly:

> But on the other hand, it is also necessary not to dwell [too much] on the mathematical [parts of the dialogue] as some do. For this too engenders false opinions in the audience in as much as they come to think of physical figures being the same thing as mathematical numbers.[10]

Similarly, in Section 3.7.3 we noted that William of Conches emphasized that there are multiple ways in which numbers may refer to things: magnitude, similarity, or order. Mathematics provides the tools for describing nature, or, in more modern language, for modeling natural phenomena. How and why this works has long been a matter for philosophical debate. Some might ascribe it to nothing more than our ability to project the structures of our interior thought onto the exterior world; others, such as Plato or the author or the Book of Wisdom, ascribe it to the mathematical intent of a creator. Even Dirac, in the talk referenced above, sees mathematical design as an innate quality of nature:

> There is no logical reason why the second method [mathematical reasoning] should be possible at all, but one has found in practice that it does work and meets with remarkable success. This must be ascribed to some *mathematical quality* in Nature, a quality which the casual observer of Nature would not suspect, but which nevertheless plays an important role in Nature's scheme.[11]

Yet both Proclus and William of Conches remind us that we must not mistake the model for the reality. Moreover, care must be taken to keep in mind exactly what is being modeled, whether, for example, it be the geometric shape of a some physical entitity or some other property of that object. For an example from modern physics, at times we might think of electrons as behaving like particles and at other times like waves. But these are both just mathematical models, as electrons are neither.

Another consideration is the distinction between prediction and explanation. Does a particular mathematical model of nature seek to explain why things are as they are, or does it only aim to predict past and future behavior? Kepler's laws of planetary motion, for example, provided very accurate predictions of the paths of the planets, yet did not have any explanatory power. Subsequently, Newton's law of universal gravitation provided an explanation for Kepler's laws, and hence for planetary motion, although the nature of gravity itself is still not understood. Aristotelian science was more often centered on searching for causes—that is, explanations—than it was on predictive power. Hence, for example, the classical

10 Proclus, *Commentary on Plato's Timaeus*, 69.

11 Dirac, "The Relation Between Mathematics and Physics," 122.

Greeks looked to explain the motion of the planets through superimposing circular motions, thinking that circular motion was fundamental to nature. This led to more and more complex models for the solar system until Kepler replaced explanation with description. The neoplatonic description of the elements is in a somewhat similar (although much simpler) position, providing a model with no predictive power while attempting to explain why a three-dimensional world would be built from exactly four elements.

9.2 Scientific Description or Numerology?

Loosely speaking, numerology involves drawing a connection between a number or a sequence of numbers, or, more generally, some function or mathematical model, and an observed arithmetical or physical pattern, without any underlying intrinsic mathematical or physical theory. For example, that 153 is the sum of the cubes of its digits, namely,

$$153 = 1^3 + 5^3 + 3^3,$$

is a bit of mathematical numerology because it says nothing which is essential to the number 153, but only something about the particular representation it has in the Indo-Arabic system. Indeed, the statement is not true in other base systems[12] and does not even make sense if, for example, we denoted the number in Roman numerals as CLIII. As an example from popular culture, in the 1998 movie *Pi*, a mathematician searches for a 216 (= $6\times6\times6$) digit number thought to have both mystical and financial powers.[13] Of course, one would not expect a number with such significance, especially powers tied to the Torah, to be specified by the number of digits it has in a paticular numeration system, least of all the Indo-Arabic system.

The Titius-Bode law provides an example of possible numerology from the study of the natural world. This putative law states that, in astronomical units,[14] the distance d_n from the sun to the nth planet satisfies

$$d_1 = 0.4, d_n = 0.4 + 0.3 \cdot 2^{n-2}, n = 2, 3, 4, \dots.$$

Thus, the law states that, for example, Mercury is 0.4 au from the Sun, Venus is 0.7 au, and the earth is, of course, 1.0 au. Although not backed by any physical or mathematical theory when first proposed in the 18th century, the predicted distances

12 For example, in binary the number is written as 10011001, so the sum of the cubes of the digits is 4, or 100 in binary.

13 *Pi*, dir. Darren Aronofsky, prod. Eric Watson and Scott Vogel (Artisan Entertainment, 1998).

14 One astronomical unit, denoted au, is equal to the average distance of the earth from the sun.

were close to the actual distances for the planets known at the time, and continued to be so for the next two planetary discoveries. However, the law did not hold for Neptune when it was first identified in 1846.

Still, the line between speculative mathematics and numerology can be hard to draw. Finding a single curve out of a collection of curves that in some way best fits a set of data is often an important step in understanding natural phenomenon and in predicting its future behavior. Kepler's fitting an ellipse to the observed positions of Mars to predict the orbit of Mars is but one famous example of finding a curve which best represents a set of data.

The philosopher and statistician I. J. Good has argued that a mathematical model should not be ignored simply because no matching scientific model has yet been found. For Good, "a physical theory is 'numerological' if it provides a formula for the calculation of physical quantities with little or no attempt to explain the formula."[15] He argues that the determination as to whether a certain theory is a true scientific theory, or just a numerological one, requires a combination of scientific and satistical reasoning, as "all science is numerological to some extent" and "[a]ll judgments are subjective," with statistics serving "[t]o increase the objectivity of judgments."[16] Although initially addressing the use of statistics to help understand what the scientific attitude should be towards the Titius-Bode law, in replying to comments about his argument, Good cites the example of Johann Balmer's discovery of a number sequence modeling the frequencies of the spectrum lines of hydrogen. Proposed by Balmer in the latter part of the 19th century, no explanatory physical counterpart underpinning this sequence existed until Niels Bohr provided one, with quantum mechanics, many years later. Good asks,

> Should the referee of Balmer's paper have rejected the paper on the grounds that no scientific model was put forward for the formula? Obviously no, because the formula was such a good fit that it was convincing that there must exist some physical explanation for it.
> The question is whether a piece of scientific numerology unsupported by a model is sufficiently striking to make us say that people ought to look for a scientific model in order to explain it.[17]

Again, we see that identifying what is numerology and what is sound mathematical or physical thinking is not always straighforward. Moreover, drawing the distinction may in part depend on the mathematical and observational tools at hand.

15 I. J. Good, "A Subjective Evalutation of Bode's Law and an 'Objective' Test for Approximate Numerical Rationality," *Journal of the American Statistical Association* 64, no. 325 (March 1969): 28.
16 Good, "A Subjective Evaluation of Bode's Law," 28.
17 F. J. Anscombe et al., "Discussion of Paper by I. J. Good," *Journal of the American Statistical Association* 64, no. 325 (March 1969): 65.

Does Plato's description of the elements in the *Timaeus*, subsequently developed by the neoplatonists throughout the Middle Ages, stay within the bounds of scientific inquiry, or does it cross the line into numerology? The answer may be clear from a modern perspective, but not so clear in the past. Certainly, the explanation seems too simplistic and ad hoc to be the basis for a serious representation of the structure of matter in the world. Nevertheless, this description is an attempt to use the observed three-dimensionality of the world to model how we might explain the types of matter our senses encounter in nature.

9.3 Elements or States of Matter?

To this point, we have identified the ancient notion of the four elements with the modern conception of the more than ninety naturally occurring chemical elements, basic building blocks of the physical world which cannot be broken down into simpler substances. However, this identification may be somewhat simplistic and is certainly anachronistic. In fact, the medieval conception of the composition of matter was multilayered and far from monolithic. In particular, philosophers and natural scientists of the time debated the nature of the matter of the physical world in terms of *prima materia*, the *elementum*, and the *elementatum*. The first refers to matter without differentiation or form, the second to the elements in their pure form, and the third to the elements as we encounter them with our senses. For example, William of Conches, in his gloss of the *Timaeus*, says:

> [Q]uantum ad materias: ile primum optinet locum, pura elementa secundum, ista visibilia elementa tercium.[18]

> [So far as concerning material things: first they occupy a place, next they are pure elements, third they are visible elements.]

In other words, prime matter creates space, but does not have form. The pure elements are the familiar four elements as they exist as particles, but are not perceptible to our senses. Finally, the matter we encounter with our senses, having added to the *elementa* the sensible qualities of the Aristotelian contraries dry/wet and cold/hot, is a mixture of these unsensed particles of pure form. But opinions

---

18 Guillelmi de Conches, *Guillelmi de Conchis: Glosae super Platonem*, CLVIII, 286.

differed among the various authors who wrote on the subject, with William's own view changing over time.[19]

In this Aristotelian framework, each element has one quality natural to itself and one quality shared with another element. So earth is dry of itself, water is cold of itself, air is wet of itself, and fire is hot of itself; earth shares coldness with water, water shares wetness with air, air shares hotness with fire, and fire share dryness with earth.[20] When two elements share a quality, and the other quality of one is transformed to the contrary quality of the other, then one element transforms into the other:

> Aqua enim unam habet transmutationem in terram. Cum enim frigiditatis et siccitatis <sit> constringere et solidum facere, nichil vero sit solidius aqua preter terram, aqua spissata transit in terram.[21]

> [For water has a single transmutation into earth. For since the result of coldness and dryness is to constrain and make something solid, and nothing is more solid than earth, thickened water changes into earth.]

When stated in this manner, the *elementa* sound more like states of matter than like elementary constituents of matter. In some approximation, it may be that a closer modern equivalent to the ancient notion of four elements is the concept of the states of matter. There are four common observable states of matter: solid, liquid, gas, and plasma. It has been suggested that four elements of Empedocles correspond to these states, as in their common forms, earth is solid, water is liquid, air is gas, and fire (if sufficiently hot, as is the case for the fire of the sun) is plasma.[22] Hence we may view the neoplatonic description of the elements given in the *Algorismus* as an attempt to explain the four states of matter in terms of the three dimensions of physical space. In this way, the four elements are identified more directly with their qualities than with their constituent particles.

With this identification, the transition from one state of matter to another, such as from solid to liquid (as when ice melts), is modeled in the neoplatonic theory geometrically as a transformation of one of the dimensions of a rectangular box. Thus earth, represented as $8 = 2 \times 2 \times 2$, transforms into water, represented as

---

19 See, for example: Irene Caiazzo, "The Four Elements in the Work of William of Conches," in *Guillaume de Conches: Philosophie et science au XIIe siècle*, trans. Michael Chase (Florence: Sismel – Edizioni del Galluzzo, 2011), 3–66.

20 Guillelmi de Conches, *Guillelmi de Conchis: Glosae super Platonem*, CLVIII, 285.

21 Guillelmi de Conches, *Guillelmi de Conchis: Glosae super Platonem*, CLXII, 294.

22 For example, see: Mitsuru Kikuchi, *Frontiers in Fusion Research: Physics and Fusion* (London: Springer-Verlag, 2011), 12.

$12 = 2 \times 2 \times 3$. Two changing to three is a discrete jump, passing from one integer directly to the next. This would be similar to our understanding of the elements in the periodic table, where the change from one element to the next is a discrete process of incrementing the atomic number. However, the change from one state of matter to the next, as in the change from solid to liquid when ice melts, is a continuous process which, for example, may be achieved by a continuous change in temperature. This too is compatible with the geometric view presented in the *Algorismus*: one can certainly picture, as in figure 3.2, one side of the cube representing earth (solid) stretching (melting) until it represents water (liquid). Is this a useful model? Perhaps not from our current perspective. We prefer models that are predictive as well as explanatory. Yet we have always sought pictures or other types of models that might impart some insight into the physical behavior of the world around us. As but one example, when trying to understand how the universe could be finite yet unbounded and continually expanding, we might picture the universe as the two-dimensional surface of a balloon as it inflates. This model is in many ways misleading, yet at the some time provides us with some insight into what is, of itself, beyond our ability to imagine. In the end, we can't truly capture the complexity of the world either in pictures or mathematics, but that should not stop us from trying to illuminate as much as we can with whatever tools and language we have available.

10 Conclusion

The Indian calculus of Muḥammed ibn Mūsā al-Khwārizmī, which spread across Western Europe in the 12th through 14th centuries in both Latin and vernacular versions, did not represent a major development in mathematics. Indeed, mathematics, of itself, is concerned with the intrinsic relationships amongst numbers, not in the means of representing those numbers and the methods for computing with them. For example, that 13 is a prime number and 28 is a perfect number are interesting mathematical facts determined not by how the numbers are represented, but by how they are related to other integers.

However, the Indian calculus did represent a major development in the field of algorithmics. The importance of the work of al-Khwārizmī and its numerous descendants lies in the methods—that is, the algorithms—it describes. These algorithms allow us to encode numbers in a certain form which is then amenable to specific procedures for desired computational results. These methods differ from algorithms specified in a modern computer language only in the way in which they are implemented—that is (for example) on dust boards rather than in the electronics of digital computers. The Indian calculus tells us, step-by-step, how to encode and input the data, how to keep track of intermediate results, and how to display and even error check the final result. Along the way, we may even encounter branches in the code, as, for example, when the square root algorithm tells us to do one thing if the number has an odd number of digits and something different if it has an even number of digits or when the subtraction algorithm tells us how to proceed differently when we encounter a cipher.

The algorithms of the short text on the geometry of a circle are of a much simpler sort than those of the *Algorismus*. Moreover, they are not dependent on the particular representation given to the numbers involved, although the actual computations indicated might require the algorithms of the *Algorismus* to complete. The instructions describe how to find some length given information about another. They are essentially one-step algorithms, but algorithms nevertheless. The intent in this section on geometry is not to present some novel mathematics or even to explain the geometry behind the computations; the text is simply a practical guide on how to make some useful computations. As with the *Algorismus* itself, the text falls into the area which Knuth would describe as algorithmics.

The point here is that, looking back from our current vantage point, we should not categorize these texts within the history of mathematics. Rather, they fall within the historical development of computational methods, or algorithmics, or, in modern English usage, computer science. But the latter term is a misnomer, as we now identify the term 'computer' with electronic digital computational machines. Not

https://doi.org/10.1515/9781501522376-010

only does the word itself predate the development of the modern machines,[1] but the science itself does not require the existence of the machines. Knuth has tried to reclaim this history for his discipline, noting that "[o]ne of the ways to help make computer science respectable is to show that it is deeply rooted in history, not just a short-lived phenomenon."[2] Looking at these texts as examples in the history of algorithmics shows them in a different light than when they are considered within the history of mathematics. Whereas they are minor works in terms of the development of mathematics, they (in particular, the *Algorismus*) are important examples in the development of algorithmics.

Much of the rest of the manuscript GKS 1812 4to is devoted to applied sciences such as astronomy and *computus*. Thus, what differentiates these two texts from the rest of the manuscript is not so much their mathematical nature but their emphasis on practical algorithms for performing the sorts of arithmetical and geometrical calculations that underlie the practical sciences treated elsewhere. Although the objects of study in these sections are mathematical—numbers in the case of the *Algorismus* and geometric figures in the other text—the interest is not in general properties of the objects themselves but in the procedures for working with them. On the other hand, the objects of study in the astronomical and calendrical parts of the manuscript are not mathematical at all, but are physical objects—in particular, the positions and movements of the fixed stars and the sun and the moon. Yet one cannot begin to understand the properties and motions of these physical objects without first knowing how to work with the numbers and geometric figures with which we model their behavior.

In this context, the final section of the Old Norse *Algorismus* is notable for its change in tone and emphasis. Here, the interest lies not in rules for working with numbers with a particular representation, but in the relationship of geometric proportion between numbers. This relationship is an intrinsic property of the numbers themselves, independent of how one might choose to write them. Moreover, the discussion fits into a long history of attempts to explain the nature of the world in terms of relationships between numbers. Simplistic as this connection might seem compared to modern mathematics and physics, it is nevertheless true that a similar search continues to this day as we look for deep connections between the pat-

1 In the earliest usage cited by the *Oxford English Dictionary*, from 1613, a computer is "a person who makes calculations or computations." See: *"computer (n.)"*, in *Oxford English Dictionary* (Oxford: Oxford University Press, 2025), accessed December 10, 2025, https://doi.org/10.1093/OED/9413010810.

2 Donald E. Knuth, "Ancient Babylonian Algorithms," *Communications of the ACM* 15, no. 7 (1972): 671.

terns found in numbers and the modern description of the world found in quantum physics.

In his *apologia* for his life, the English mathematician G. H. Hardy addresses the nature of lasting fame for a mathematician. He claims that "as history proves abundantly, mathematical achievement, whatever its intrinsic worth, is the most enduring of all"[3] and, as an example,

> Greek mathematics is 'permanent,' more permanent even than Greek literature. Archimedes will be remembered when Aeschylus is forgotten, because languages die and mathematical ideas do not. 'Immortality' may be a silly word, but probably a mathematician has the best chance of whatever it may mean.[4]

Hardy admits that this fame may be somewhat anonymous, as we can see with the case of al-Khwārizmī. Although his name is deeply ingrained in our language, through words such as *algorismus* and algorithm, few associate the words with the man himself, or even realize that the words are connected with any particular person at all. Yet Hardy would argue that this is precisely the form of enduring renown that a mathematician would choose.[5]

3 G. H. Hardy, *A Mathematician's Apology* (Cambridge: Cambridge University Press, 2012), 80.
4 Hardy, *A Mathematician's Apology*, 81.
5 Hardy, *A Mathematician's Apology*, 153.

References

Primary Sources – Old Norse

"AM 415 4to." In *Handrit.is*. Reykjavík: The Árni Magnússon Institute for Icelandic Studies. Accessed March 2, 2026. https://handrit.is/manuscript/view/is/AM04-0415.

AM 544 4to. Copenhagen: Institut for Nordiske Studier og Sprogvidenskab. Accessed December 10, 2025. https://sprogsamlinger.ku.dk/q.php?p=ds/hjem/mapper/9198.

AM 685 d 4to. Copenhagen: Institut for Nordiske Studier og Sprogvidenskab. Accessed December 10, 2025. https://sprogsamlinger.ku.dk/q.php?p=ds/hjem/mapper/11658.

AM 736 III 4to. Copenhagen: Institut for Nordiske Studier og Sprogvidenskab. Accessed December 10, 2025. https://sprogsamlinger.ku.dk/q.php?p=ds/hjem/mapper/12327.

"AM 736 III 4to." In *Handrit.is*. Copenhagen: Arnamagnæan Institute. Accessed March 2, 2026. https://handrit.is/manuscript/view/da/AM04-0736-III.

GKS 1812 4to. Copenhagen: Institut for Nordiske Studier og Sprogvidenskab. Accessed December 10, 2025. https://sprogsamlinger.ku.dk/q.php?p=ds/hjem/mapper/35612.

"GKS 1812 4to." In *Handrit.is*. Reykjavík: The Árni Magnússon Institute for Icelandic Studies. Accessed December 10, 2025. https://handrit.is/en/manuscript/view/is/GKS04-1812.

Hauksbók. Edited by Finnur Jónsson and Eiríkur Jónsson. Copenhagen: Thieles Bogtrykkeri, 1892–96.

Primary Sources – Latin

"Adalbaldi, Episcopi Trajectensis, opusculum super illud Boëtii: O qui perpetua mundum ratione gubernas." In BNF Latin 7361, 46v–56v. Bibliothèque nationale de France. Accessed December 10, 2025. https://archivesetmanuscrits.bnf.fr/ark:/12148/cc665714.

Adalbold of Utrecht. "O qui perpetua: Der Kommentar des Adalbold von Utrecht." In *Serta Mediaevalia: Textus varii saeculorum X–XIII*, edited by R. B. C. Huygens, 121–40. Turnhout: Brepols, 2000.

Adam von Bremen. *Hamburgische Kirchengeschichte*. Edited by Bernhard Schmeidler. Hanover and Leipzig: Hahnsche Buchhandlung, 1917.

"Alexander <de Villa Dei>: Carmen de Algorismo." In UB Erfurt, Dep. Erf. CA. 12° 017, 68va–70va. Universitätsbibliothek Erfurt – Handschriften der Amploniana. Accessed September 13, 2023. https://dhb.thulb.uni-jena.de/receive/ufb_cbu_00023580.

Alexander de Villa Dei. "Carmen de algorismo." In BNF Latin 7475. Bibliothèque nationale de France. Accessed December 10, 2025. https://archivesetmanuscrits.bnf.fr/ark:/12148/cc94612b.

Alexander de Villedieu. "Carmen de algorismo." In *Rara Mathematica*, edited by James Orchard Halliwell, 73–83. London: John William Parker, 1839.

"Algorismus sive ars introductorius in numerum." In Kgl. bibl. X 767, 15r–23v. Kungliga biblioteket, Stockholm.

Allard, André, ed. *Muhammad Ibn Mūsā Al-Khwārizmī: Le Calcul Indien (Algorismus)*. Namur, Belgium: Société des Études Classiques, 1992.

Ancien testament: ms. II 1639. KBR: Bibliothèque royale de Belgique. Accessed December 10, 2025. https://opac.kbr.be/LIBRARY/doc/SYRACUSE/10749304.

Baker, Peter S., and Michael Lapidge, eds. *Byrhtferth's Enchiridion*. Oxford: Oxford University Press, 1995.

Bede. "De natura rerum." In *Operum Venerabilis Bedae presbyteri*, edited by Johann Herwagen, II:1–49. Basel: Hervagius, 1563.

Bede. "De natura rerum." In *Beda Venerabilis: Opera didascalica*, edited by Charles W. Jones, I:173–234. Brepols, 1975.

Bernard of Chartres. *The Glosae Super Platonem of Bernard of Chartres*. Edited by Paul Edward Dutton. Toronto: Pontifical Institute of Mediaeval Studies, 1991.

Biblia Sacra Iuxta Vulgatam Versionem. Edited by Robert Weber and Roger Gryson. Stuttgart: Deutsche Bibelgesellschaft, 2007.

Böckerman, Robin Wahlsten, ed. "Bavarian B, clm 14482c." In *The Bavarian Commentary and Ovid*, 333–65. Cambridge: Open Book Publishers, 2020.

Böckerman, Robin Wahlsten, ed. "Edition and Translation of clm 4610." In *The Bavarian Commentary and Ovid*, 190–321. Cambridge: Open Book Publishers, 2020.

Böckerman, Robin Wahlsten, ed. *The Bavarian Commentary and Ovid*. Cambridge: Open Book Publishers, 2020.

Boethius. *Boèce: Institution arithmétique*. Edited by Jean-Yves Guillaumin. Paris: Les Belles Lettres, 1995.

Boethius. "De consolatione philosophiae." In *De consolatione philosophiae. Opuscula theogica*, edited by Claudio Moreschini, 1–162. Munich and Leipzig: K. G. Saur, 2005.

Boncompagni, Baldassarre, ed. "Ionnis Hispalensis: Liber Algorismi de pratica arismetrice." In *Trattati d'aritmetica*, vol. II. Rome: Tipografia Delle Scienze Fisiche e Matematiche, 1857.

Byrhtferth. "Epilogus." In St John's College MS 17, 12v–13r. Oxford: Bodleian Library, University of Oxford. Accessed December 10, 2025. https://digital.bodleian.ox.ac.uk/objects/cca30c56-0751-4f52-a952-bbffcb7b64e9/.

Calcidius. *Calcidius: Commentaire au Timée de Platon*. Edited by Béatrice Bakhouche. Translated by Béatrice Bakhouche and Luc Brissson. 2 vols. Paris: Librairie Philosophique J. Vrin, 2011.

Calcidius. *On Plato's Timaeus*. Edited and translated by John Magee. Cambridge, MA: Harvard University Press, 2016.

Calcidius. *Timaeus: A Calcidio translatus commentarioque instructus*. Edited by J. H. Waszink. London and Leiden: Aedibus Instituti Warburgiana et E. J Brill, 1975.

Cambridge University Library, MS Ff.6.13. Cambridge: Cambridge University Library.

Cantor, Moritz. "Über einen Codex des Klosters Salem." *Zeitschrift für Mathematik und Physik*, no. 10 (1865): 1–16.

"*Chalcidii* interpretatio et commentarius in Timaeum Platonis." In BNF Latin 6281. Bibliothèque nationale de France. Accessed December 9, 2025. https://archivesetmanuscrits.bnf.fr/ark:/12148/cc65354z.

Clarembald of Arras. "Epistola ad Dominam." In *Life and Works of Clarembald of Arras*, edited by Nikolaus M. Häring, 225–26. Toronto: Pontifical Institute of Mediaeval Studies, 1965.

Clarembald of Arras. "Expositio super librum Boetii De Hebdomadibus." In *Life and Works of Clarembald of Arras*, edited by Nikolaus M. Häring, 189–221. Toronto: Pontifical Institute of Mediaeval Studies, 1965.

Clarembald of Arras. "Tractatulus super librum Genesis." In *Life and Works of Clarembald of Arras*, edited by Nikolaus M. Häring, 226–49. Toronto: Pontifical Institute of Mediaeval Studies, 1965.

Codex Latinus Monacensis 14482. Munich: Bayerische Staatsbibliothek. Accessed December 10, 2025. https://www.digitale-sammlungen.de/en/view/bsb00046312.

"Commentarius in Timaeum Platonis." In Universitetsbiblioteket MS. C 620. Uppsala: Uppsala Universitetsbibliotek.

Conventi Soppressi E.VIII 1398. Florence: Biblioteca Nazionale Centrale di Firenze.

"De natura rerum." In Staatsbibliothek Bamberg Msc.Nat.1. Staatsbibliothek Bamberg. Accessed December 10, 2025. https://nbn-resolving.org/urn:nbn:de:bvb:22-dtl-0000025375.
Gerbert of Aurillac. "Epistola ad Adalboldum." In *Oeuvres de Gerbert*, edited by Alexandre Olleris, 477–78. Paris: Dumoulin, 1867.
Gerbert of Aurillac. "Geometria Gerberti." In *Oeuvres de Gerbert*, edited by Alexandre Olleris, 403–70. Paris: Dumoulin, 1867.
Guillelmi de Conches. *Dragmaticon philosophiae*. Edited by Italo Ronca. Turnhout: Brepols, 1997.
Guillelmi de Conches. *Glosae super Boetium*. Edited by Lodi Nauta. Turnhout: Brepols, 1999.
Guillelmi de Conches. *Guillelmi de Conchis: Glosae super Platonem*. Edited by Édouard A. Jeauneau. Turnhout: Brepols, 2006.
Halliwell, James Orchard, ed. "A Treatise on the Numeration of Algorism." In *Rara Mathematica*, 29–31. London: John William Parker, 1839.
Huygens, R. B. C., ed. "Serta Mediaevalia: Textus varii saeculorum X-XIII." Turnhout: Brepols, 2000.
Ioannis Saresberiensis. *Metalogicon*. Edited by J. B. Hall. Turnhout: Brepols, 1991.
Iohannis de Sacro Bosco. "Algorismus." In *Petri Philomenae de Dacia et Petri de S. Audomaro: Opera quadrivialia*, edited by Fridericus Saaby Pedersen, 1:165–201. Copenhagen: Societas Linguae & Litterarum Danicarum, 1983.
Iohannis de Sacro Bosco. "Tractatus de arte numerandi." In *Rara Mathematica*, edited by James Orchard Halliwell, 1–26. London: John William Parker, 1839.
Iohannis de Sacro Bosco. "Tractatus de sphera." In *The Sphere of Sacrobosco and Its Commentators*, edited by Lynn Thorndike, 76–117. Chicago: The University of Chicago Press, 1949.
Isidore de Seville. *Traité de la nature*. Edited and translated by Jacques Fontaine. Bordeaux: Féret et Fils, Éditeurs, 1960.
"*Isidori Junioris* liber de natura rerum: initium desideratur." In BNF Latin 6413. Bibliothèque nationale de France. Accessed December 10, 2025. https://archivesetmanuscrits.bnf.fr/ark:/12148/cc65498x.
Lafleur, Claude, ed. "Philosophica disciplina." In *Quatre introductions à la philosophie au XIIIe siècle*, 255–93. Montreal: Institut d'Études Médiévales, 1988.
Macrobius. *Commentarii in Ciceronis somnium Scipionis*. Edited by Ludwig von Jan. Quedlinburg and Leipzig: Godofredius Bassius, 1848.
Macrobius. *Saturnaliorum*. Edited by Ludwig von Jan. Quedlinburg and Leipzig: Godofredius Bassius, 1852.
"*Magistri Guillelmi de Conchis* opus inscriptum: Dragmaticon philosophiae." In *BNF Latin 6415*. Bibliothèque nationale de France. Accessed March 7, 2026. https://archivesetmanuscrits.bnf.fr/ark:/12148/cc655009?collect.
"MS. Ashmole 399, 64r–71r." In *Digital Bodleian*. Bodleian Library, University of Oxford. Accessed December 10, 2025. https://digital.bodleian.ox.ac.uk/inquire/p/15a20263-f647-4913-9089-a9efbb37225c.
"MS. Auct. F. 5. 29." In *Medieval Manuscripts in Oxford Libraries*. Bodleian Library, University of Oxford. Accessed December 10, 2025. https://medieval.bodleian.ox.ac.uk/catalog/manuscript_700.
"MS. Bodleian 177, 45." In *Medieval Manuscripts in Oxford Libraries*. Bodleian Library, University of Oxford. Accessed December 10, 2025. https://medieval.bodleian.ox.ac.uk/catalog/manuscript_1234.
NKS 218 4to: Macrobius, Commentarii in somnium Scipionis. Det Kongelige Bibliotek. Accessed December 9, 2025. https://permalink.kb.dk/permalink/2006/manus/33/eng/.
Ovid. *Metamorphoses*. Edited by R. J. Tarrant. Oxford: Oxford University Press, 2004.
Pal. lat. 1341. Biblioteca Apostolica Vaticana, Digivatlib. Accessed February 23, 2026. https://digi.vatlib.it/view/bav_pal_lat_1341/0196.

Pedersen, Fridericus Saaby, ed. *Petri Philomenae de Dacia et Petri de S. Audomaro: Opera quadrivialia*. Vol. 1. Copenhagen: Societas Linguae & Litterarum Danicarum, 1983.
Petrus de Dacia. "Expositio super Algorismum." In *Petri Philomenae de Dacia et Petri de S. Audomaro: Opera quadrivialia*, edited by Fridericus Saaby Pedersen, 1:69–163. Copenhagen: Societas Linguae & Litterarum Danicarum, 1983.
Schmid, Toni, ed. "Ein Timaioskommentar in Sigtuna." *Classica et Mediaevalia: Revue Danoise de Philologie et d'Histoire* X, no. 2 (1949): 220–266.
"Summa super naturalibus quaestionibus." In Bodleian Library MS. Digby 107. Bodleian Library, University of Oxford. Accessed December 10, 2025. https://digital.bodleian.ox.ac.uk/objects/71deba7a-66d2-4d92-8647-25421da56794/surfaces/e54e853c-b91c-48f8-9f5a-d3e9610d7a84/.
Thierry of Chartres. *The Commentary on the De arithmetica of Boethius*. Edited by Irene Caiazzo. Toronto: Pontifical Institute of Mediaeval Studies, 2015.
Thomas Bradwardine. *Arithmetica speculativa*. Paris: Guy Marchand, 1496.
"Timaeus Platonis." In MS Brussels 9625–26. KBR: Bibliothèque royale de Belgique. Accessed March 2, 2026. https://uurl.kbr.be/bib/17367397.
"Timaeus *Platonis*, de graeco in latinum, petente Josio, sive potiùs Osio, Cordubensi Episcopo, à Chalcidio, viro claro translatus, et miro ingenio commentatus et elucidatus." In BNF Latin 6282. Bibliothèque nationale de France. Accessed December 9, 2025. https://archivesetmanuscrits.bnf.fr/ark:/12148/cc653556.
"Varia de algorismo et computo." In BNF Nouvelles acquisitions latines 657. Bibliothèque nationale de France. Accessed March 1, 2026. https://archivesetmanuscrits.bnf.fr/pageCollections.html?col=1.

Primary Sources – French

Karpinski, Louis C. "A Thirteenth Century Algorism in French Verse." *Isis* 11, no. 1 (1928): 45–84.
Mortet, Victor. "Le plus ancien traité français d'algorisme." *Bibliotheca mathematica* 3, no. 9 (1908): 55–64.

Primary Sources – English

Adam of Bremen. *History of the Archbishops of Hamburg-Bremen*. Translated by Francis J. Tschan. New York: Columbia University Press, 2002.
Alcinous. *The Handbook of Platonism*. Translated by John Dillon. Oxford: Clarendon Press, 1993.
Aristotle. "Metaphysics." In *The Complete Works of Aristotle: The Revised Oxford Translation*, edited by Jonathan Barnes, translated by W. D. Rouse, vol. 2. Princeton, New Jersey: Princeton University Press, 1984.
Aristotle. "On Generation and Corruption." In *The Complete Works of Aristotle: The Revised Oxford Translation*, edited by Jonathan Barnes, translated by H. H. Joachim, vol. 1. Princeton, New Jersey: Princeton University Press, 1984.
Bede. *Bede On the Nature of Things and On Times*. Edited by Calvin B. Kendall and Faith Walllis. Translated by Calvin B. Kendall and Faith Wallis. Liverpool: Liverpool University Press, 2010.
Boethius. *Boethian Number Theory: A Translation of the De Insitutione Arithmetica*. Translated by Michael Masi. Amsterdam: Rodopi, 1983.

Conches, Willliam of. *A Dialogue on Natural Philosophy*. Translated by Italo Ronca and Matthew Curr. Notre Dame, IN: University of Notre Dame Press, 1997.

Crossley, John N., and Alan S. Henry. "Thus Spake al-Khwārizmī: A Translation of the Text of Cambridge University Library Ms. Ii.vi.5." *Historia Mathematica* 17 (1990): 103–31.

Euclid. *The Thirteen Books of Euclid's Elements*. Translated by T. L. Heath. Vol. 2. Cambridge: Cambridge University Press, 1908.

Euclid. *The Thirteen Books of Euclid's Elements*. Translated by T. L. Heath. Vol. 3. Cambridge: Cambridge University Press, 1908.

Guthrie, Kenneth Sylvan, ed. and trans. "On the Nature of the Universe." In *An Anthology of Ancient Writings Which Relate to Pythagoras and Pythagoroean Philosophy*, 203–11. Grand Rapids, MI: Phanes Press, 1987.

Isidore of Seville. *On the Nature of Things*. Edited and translated by Calvin B. Kendall and Faith Wallis. Liverpool: Liverpool University Press, 2016.

John of Sacrobosco. "The Sphere of Sacrobosco." In *The Sphere of Sacrobosco and Its Commentators*, edited and translated by Lynn Thorndike, 118–42. Chicago: The University of Chicago Press, 1949.

John of Salisbury. *Metalogicon*. Edited and translated by J. B. Hall. Turnhout: Brepols, 2013.

Karpinski, Louis C. "Two Twelfth Century Algorisms." *Isis* 3, no. 3 (1921): 396–413.

Laks, André, and Glenn W. Most, eds. *Early Greek Philosophy*. Vol. V. Cambridge, MA: Harvard University Press, 2016.

Leonardo Pisano. *Fibonacci's Liber Abaci*. Translated by L. E. Sigler. New York: Springer, 2002.

Macrobius. *Commentary on the Dream of Scipio*. Translated by William Harris Stahl. New York: Columbia University Press, 1990.

Nicomachus. *Introduction to Arithmetic*. Edited by Martin Luther D'Ooge. New York: The Macmillan Company, 1926.

Plato. *Plato: Timaeus and Critias*. Translated by A. E. Taylor. London: Methuen & Co. Ltd., 1929.

Proclus. *A Commentary on the the First Book of Euclid's Elements*. Translated by Glenn R. Morrow. Princeton, New Jersey: Princeton Universy Press, 1970.

Proclus. *Commentary on Plato's Timaeus*. Edited and translated by Harold Tarrant. Vol. I. Cambridge: Cambridge University Press, 2007.

Proclus. *Commentary on Plato's Timaeus*. Edited and translated by Dirk Baltzly. Vol. III. Cambridge: Cambridge University Press, 2007.

Steele, Robert, ed. "A Treatise on the Numeration of Algorism." In *The Earliest Arithmetics in English*, 70–71. London: Oxford University Press, 1922.

Steele, Robert, ed. "Carmen de algorismo." In *The Earliest Arithmetics in English*, 72–80. London: Oxford University Press, 1922.

Steele, Robert, ed. "The Crafte of Nombrynge." In *The Earliest Arithmetics in English*, 3–32. London: Oxford University Press, 1922.

Steele, Robert, ed. *The Earliest Arithmetics in English*. London: Oxford University Press, 1922.

Primary Sources - Greek

Alcinoos. *Alcinoos: Enseignement des doctrines de Platon*. Edited by John Whittaker. Translated by Pierre Louis. Paris: Les Belles Lettres, 1990.

Euclid. "In Euclidis elementorum libri." In Manuscript Gr. 190.pt.2, 175r–247v, 283r–292v. Vatican: Vatican Library, Digivatlib. Accessed February 11, 2026. https://digi.vatlib.it/mss/detail/Vat.gr.190.pt.2.
Euclid. "In Euclidis elementorum libri." In Manuscript Gr. 190.pt.1, 3r–174v. Vatican: Vatican Library, Digivatlib. Accessed February 11, 2026. https://digi.vatlib.it/mss/detail/Vat.gr.190.pt.1.
Iamblichus. *De vita pythagorica*. Edited by Ludovicus Deubner. Stuttgart: B. G. Teubner, 1975.
Proclus. *In Platonis Timaeum commentaria*. Edited by Ernst Diehl. Leipzig: B. G. Teubner, 1904.

Secondary Sources

Allard, André. "The Influence of Arabic Mathematics in the Medieval West." Chap. 16 in *Encyclopedia of the History of Arabic Science*, edited by Roshdi Rashed, 2:514–55. London: Taylor & Francis, 1996.
Ambrosetti, Nadia. "Algorithmic in the 12th Century: The Carmen de Algorismo by Alexander de Villa Dei." In *History and Philosophy of Computing: Third International Conference, 2015*, edited by Fabio Gadduci and Mirko Tavosanis, 71–86. Cham, Switzerland: Springer, 2016.
Anscombe, F. J., Irwin Bross, H. O. Hartley, Rolf Bargmann, H. A. David, Marvin Zelen, R. L. Anderson, Miles Davis, and I. J. Good. "Discussion of Paper by I. J. Good." *Journal of the American Statistical Association* 64, no. 325 (March 1969): 50–66.
Beckman, Natanael, and Kristian Kålund, eds. *Alfræði Íslenzk*. Vol. 2. Copenhagen: S. L. Møllers Bogtrykkeri, 1916.
Bekken, Otto B. *On the CUBUS PERFECTUS of the Algorismus in Hauksbók*. Agder, Norway: Agder distriktshøgskole, 1986.
Bekken, Otto B., and Marit Christoffersen. "Algorismus." In *Medieval Scandinavia: An Encyclopedia (Encyclopedias of the Middle Ages)*, edited by Phillip Pulsiano and Kirsten Wolf, 8–9. Oxon: Routledge, 2016.
Benedict, Suzan Rose. *A Comparative Study of the Early Treatises Introducing into Europe the Hindu Art of Reckoning*. Concord, New Hampshire: The Rumford Press, 1914.
Bishop, Errett. *Foundations of Constructive Analysis*. New York: McGraw-Hill, 1967.
Björn Gunnlaugsson. *Tölvisi*. Reykjavik: Hið íslenzka bókmenntafélag, 1865.
Bober, Harry. "In Principio. Creation Before Time." In *De Artibus Opuscula XL: Essays in Honor of Erwin Panofsky*, edited by Millard Meiss, 1:13–28. New York: New York University Press, 1961.
Borel, Émile. "Le calcul des intégrales définies." *Journal de Mathématiques Pures et Appliqueées 6ᵉ série* 8 (1912): 159–210.
Brown, George Hardin. *A Companion to Bede*. Woodbridge, England: The Boydell Press, 2009.
Caiazzo, Irene. "The Four Elements in the Work of William of Conches." In *Guillaume de Conches: Philosophie et science au XIIe siècle*, translated by Michael Chase, 3–66. Florence: Sismel – Edizioni del Galluzzo, 2011.
Cameron, Alan. "The Date and Identity of Macrobius." *The Journal of Roman Studies* 56 (1966): 25–38.
Chabert, Jean-Luc, et al. *Histoire d'algorithmes: Du caillou à la puce*. Edited by Jean-Luc Chabert. Paris: Belin, 1993.
Chadwick, Henry. *The Consolations of Music, Logic, Theology, and Philosophy*. Oxford: Clarendon Press, 1981.
Chasles, Michel. *Aperçu historique sur l'origine et le développement des méthodes en géometrie*. Brussels: M. Hayes, Imprimeur de l'Académie Royale, 1837.
Clerval, Alexandre. *Les écoles de Chartres au Moyen-Age*. Chartres: Librairie R. Selleret, 1895.

Conte, Gian Biagio. *Latin Literature: A History*. Translated by Joseph B. Solodow. Baltimore and London: The Johns Hopkins University Press, 1994.

Dillon, John. *The Middle Platonists: 80 B.C. to A.D. 220*. Ithaca, New York: Cornell University Press, 1996.

Dirac, Paul Adrien Maurice. "The Relation Between Mathematics and Physics." *Proceedings of the Royal Society (Edinburgh)* 59 (1938–39): 122–29.

Dutton, Paul Edward. "The Uncovering of the Glosae Svper Platonem of Bernard of Chartres." *Mediaeval Studies* 46 (1984): 192–221.

Eastwood, Bruce S. "The Diagram of the Four Elements in the Oldest Manuscripts of Isidore's *De natura rerum*." *Studi Medievali* 42, no. 2 (2001): 547–70.

Eneström, Gustaf. "Sur quelques propositions de planimétrie énoncées dans un manuscrit norvégien du 14e siècle." *Bibliotheca Mathematica* 12 (1898): 19–22.

Etheridge, Christian. "The Context of the Astronomical Treatises in GKS 1812 4to." In *A World in Fragments: Studies on the Encyclopedic Manuscript GKS 1812 4to*, edited by Gunnar Harðarson, Christian Ethridge, Guðrún Nordal, and Svanhildur Óskarsdóttir, 227–58. Reykjavík: Stofnun Árna Magnússonar í Íslenskum Fræðum, 2021.

Folkerts, Menso. "Early Texts on Hindu-Arabic Calculation." *Science in Context* 14, nos. 1–2 (2002): 13–38.

Foulet, Alfred, and Mary Blakely Speer. *On Editing Old French Texts*. Lawrence, Kansas: The Regents Press of Kansas, 1979.

Gersh, Stephen. "One Thousand Years of Proclus: An Introduction to His Reception." In *Interpreting Proclus: From Antiquity to the Renaissance*, edited by Stephen Gersh, 1–29. Cambridge: Cambridge University Press, 2014.

Good, I. J. "A Subjective Evalutation of Bode's Law and an 'Objective' Test for Approximate Numerical Rationality." *Journal of the American Statistical Association* 64, no. 325 (March 1969): 23–49.

Gorman, Michael M. "The Diagrams in the Oldest Manuscripts of Isidore's *De natura rerum*." *Studi Medievali* 42, no. 2 (2001): 527–45.

Grant, Edward, ed. *A Source Book in Medieval Science*. Cambridge, MA: Harvard University Press, 1974.

Grimm, Richard E. "The Autobiography of Leonardo Pisano." *The Fibonacci Quarterly* 11, no. 1 (1973): 99–104.

Guðvarður Már Gunnlaugsson. "Stafrétt eða samræmt? Um fræðilegar útgáfur og notendur þeirra." *Gripla* 14 (20003): 197–235.

Gunnar Harðarson. "A *Divisio Philosophiae* in the Medieval Icelandic Manuscript GKS 1812°." *Cahiers de l'Institut du Moyen-Âge Grec et Latin*, no. 84 (2015): 2–21.

Gunnar Harðarson. "Medieval Encyclopedic Literature and Icelandic Manuscripts." In *A World in Fragments: Studies on the Encyclopedic Manuscript GKS 1812 4to*, edited by Gunnar Harðarson, Christian Ethridge, Guðrún Nordal, and Svanhildur Óskarsdóttir, 13–38. Reykjavík: Stofnun Árna Magnússonar í Íslenskum Fræðum, 2021.

Gunnar Harðarson and Stefán Karlsson. "Hauksbók." In *Medieval Scandinavia: An Encyclopedia (Encyclopedias of the Middle Ages)*, edited by Phillip Pulsiano and Kirsten Wolf, 271–72. Oxon: Routledge, 2016.

Hacking, Ian. *Why is There Philosophy of Mathematic at All?* Cambridge: Cambridge University Press, 2014.

Hankel, Hermann. *Zur Geschichte der Mathematik in Alterthum und Mittelalter*. Leipzig: B. G. Teubner, 1874.

Haraldur Bernharðsson. "Scribes and Scribal Practice in GKS1812 4to." In *A World in Fragments: Studies on the Encyclopedic Manuscript GKS 1812 4to*, edited by Gunnar Harðarson, Christian Ethridge, Guðrún Nordal, and Svanhildur Óskarsdóttir, 61–113. Reykjavík: Stofnun Árna Magnússonar í Íslenskum Fræðum, 2021.

Haraldur Bernharðsson and Odd Einar Haugen. "Levels of Text Representation." Chap. 4, 3rd ed., edited by Odd Einar Haugen. Bergen, Norway: Medieval Nordic Text Archive, 2019. Accessed December 10, 2025. https://menota.org/HB3_ch4.xml.

Hardy, G. H. *A Mathematician's Apology*. Cambridge: Cambridge University Press, 2012.

Haugen, Odd Einar. "On the Diplomatic Turn in Editorial Philology." In *The Fantastic in Old Norse/Icelandic Literature: Sagas and the British Isles*, edited by John McKinnell, David Ashurst, and Donata Kick. Durham and York: Center for Medieval / Renaissance Studies, 2006.

Hoenig, Christina. *Plato's Timaeus and the Latin Tradition*. Cambridge: Cambridge University Press, 2018.

Høyrup, Jens. *Jacopo da Firenze's Tractatus Algorismi and Early Italian Abbacus Culture*. Basel: Birkhäuser, 2007.

Høyrup, Jens. "When is the Algorithm Concept Pertinent – and When not? Thoughts About Algorithms and Paradigmatic Examples, and About Algorithmic and Non-Algorithmic Mathematical Cultures." *AIMS Mathematics* 3, no. 1 (2018): 211–32.

Jeauneau, Édouard. *Rethinking the School of Chartres*. Translated by Claude Paul Desmarais. Toronto: University of Toronto Press, 2009.

Jeauneau, Édouard A. "L'usage de la notion d'*integumentum* a travers les gloses de Guillaume de Conches." *Archives d'Histoire Doctrinale et Littéraire du Moyen Age* 23 (1957): 35–100.

Johansson, Bo Göran. "Cube Root Extraction in Medieval Mathematics." *Historia Mathematica* 38 (2011): 338–67.

Karpinski, Louis C. "The Italian Arithmetic and Algebra of Master of Jacob of Florence, 1307." *Archeion* 11, nos. 2–3 (1929): 170–77.

Kaylor, Noel Harold, Jr. "Introduction: The Times, Life, and Work of Boethius." In *A Companion to Boethius in the Middle Ages*, edited by Noel Harold Kaylor Jr. and Philip Edward Phillips, 1–46. Leiden: Brill, 2012.

Kikuchi, Mitsuru. *Frontiers in Fusion Research: Physics and Fusion*. London: Springer-Verlag, 2011.

Kingsley, K. Scarlett, and Richard Parry. "Empedocles." In *The Stanford Encyclopedia of Philosophy*, Winter 2024, edited by Edward N. Zalta. Metaphysics Research Lab, Stanford University, 2024. Accessed December 10, 2025. https://plato.stanford.edu/archives/win2024/entries/empedocles/.

Knuth, Donald E. "Algorithms in Modern Mathematics and Computer Science." In *Algorithms in Modern Mathematics and Computer Science*, edited by E. P. Ershov and D. E. Knuth, 82–99. Berlin and Heidelberg: Springer, 1981.

Knuth, Donald E. "Ancient Babylonian Algorithms." *Communications of the ACM* 15, no. 7 (1972): 671–77.

Kristín Bjarnadóttir. "Algorismus: Fornt stærðfræðirit í íslenskum handritum." *NETLA*, 2004. Accessed December 10, 2025. http://wayback.vefsafn.is/wayback/20201017180549/https://netla.hi.is/greinar/2004/001/index.htm.

Kristín Bjarnadóttir and Bjarni V. Halldórsson. "*Algorismus*: Hindu-Arabic Arithmetic in GKS 1812 4to." In *A World in Fragments: Studies on the Encyclopedic Manuscript GKS 1812 4to*, edited by Gunnar Harðarson, Christian Ethridge, Guðrún Nordal, and Svanhildur Óskarsdóttir, 171–92. Reykjavík: Stofnun Árna Magnússonar í Íslenskum Fræðum, 2021.

Kristín Bjarnadóttir and Bjarni V. Halldórsson. "The Norse Treatise Algorismus." In *Actes du 10è, colloque maghrébin sur l'histoire des mathématiques arabes*, edited by Mahdi Adbeljaouad, 67–77. Tunis: L'Association Tunisiennes des Sciences Mathématiques, 2010.

Langebek, Jacobus, ed. *Scriptores rerum danicorum medii ævi*. Vol. II. Copenhagen: Godiche, 1773.
Lapidge, Michael. "Byrhtferth of Ramsey and the 'Glossae Bridferti in Bedam'." *The Journal of Medieval Latin* 17 (2007): 384–400.
Martin, Thomas Henri. *Études sur le Timée de Platon*. Paris: Ladrange Libraire-Éditeur, 1841.
McCarthy, Cormac. *The Passenger*. New York: Alfred A. Knopf, 2022.
Moyer, Ann E. "The *Quadrivium* and the Decline of Boethian Influence." In *A Companion to Boethius in the Middle Ages*, edited by Noel Harold Kaylor Jr. and Philip Edward Phillips, 479–517. Leiden: Brill, 2012.
Munch, P. A. "Algorismus, eller Anviisning til at kjende og anvende de saakaldte arabiske Tal, efter Hr. Hauk Erlendssons Codex." In *Annaler for nordisk oldkyndighed og historie*, 353–75. Copenhagen: J. D. Qvist, 1848.
Murdoch, John E. *Album of Science: Antiquity and the Middle Ages*. New York: Charles Scribner's Sons, 1984.
Narducci, Enrico, ed. *Catalogo di manoscritti ora posseduti da Baldasssarre Boncompagni*. 2nd ed. Rome: Tipografia delle Scienze Matematiche e Fisiche, 1892.
Obrist, Barbara. *La cosmologie médiévale. Textes et images I.* Les fondements antiques. Florence: Sismel Edizioni del Galluzzo, 2004.
"ONP: Dictionary of Old Norse Prose (Manuscripts)." Department of Nordic Studies and Linguistics University of Copenhagen. Accessed December 10, 2025. https://onp.ku.dk/onp/onp.php?m.
Otisk, Marek. "The Four Elements and Their Characteristics According to the Schema in the Early Medieval Anonymous Fragment *De Quattuor Elementis*." *Cosmos and History: The Journal of Natural and Social Philosophy* 17, no. 3 (2021): 462–500.
Overgaard, Mariane. "The Organisation of the Photography and Restoration of Manuscripts Prior to Their Transfer to Iceland 1971–1996." In *Care and Conservation of Manuscripts 3. Proceedings of the Third International Seminar Held at the University of Copenhagen 14th–15th October 1996*, edited by Gillian Fellows-Jensen and Peter Springborg, translated by Gillian Fellows-Jensen, 7–27. Copenhagen: The Royal Library, 1997.
Pedersen, Olaf. "In Quest of Sacrobosco." *Journal for the History of Astronomy* 16, no. 3 (1985): 175–221.
Pedersen, Olaf. "Petrus Philomena de Dacia: A Problem of Identity." *Cahiers de l'Institut du Moyen-Âge Gred et Latin* 19 (1976): 1–54.
Pedersen, Olaf. "The Life and Work of Peter Nightingale." In *Vistas in Astronomy: New Aspects in the History and Philosophy of Astronomy*, edited by Arthur Beer, 9:3–10. Oxford: Pergamon Press, 1967.
Peirce, Benjamin. *Linear Associative Algebra*. New York: D. Van Nostrand, 1882.
Peirce, Charles S. "The Simplest Mathematics." In *Philosophy of Mathematics: Selected Writings*, edited by Matthew E. Moore. Bloomington and Indianapolis, IN: Indiana University Press, 2010.
Pi. Directed by Darren Aronofsky, produced by Eric Watson and Scott Vogel. Artisan Entertainment, 1998.
Piketty, Thomas. *Capital and Ideology*. Translated by Arthur Goldhammer. Cambridge, MA: The Belknap Press of Harvard University Press, 2020.
Rashed, Roshdi. "Algebra." Chap. 11 in *Encyclopedia of the History of Arabic Science*, edited by Roshdi Rashed, 2:324–50. London: Taylor & Francis, 1996.
Reinaud, Joseph Toussaint. "Mémoire géographique, historique et scientifique sur l'Inde, antéreieurment au milieu du XIe siécle de l'ére chrétienne, d'aprés les écrivains arabes, persans et chinois." In *Mémoires de l'Institute National de France*, 18:1–399. Paris: Académie des Inscriptions et Belles-Lettres, 1849.

Ross, Margaret Clunies, and Rudolf Simek. "Encyclopedic Literature." In *Medieval Scandinavia: An Encyclopedia (Encyclopedias of the Middle Ages)*, edited by Phillip Pulsiano and Kirsten Wolf, 164–66. Oxon: Routledge, 2016.

Russell, Josiah Cox. *Dictionary of Writers of Thirteenth Century England*. New York: Burt Franklin, 1971.

Sloughter, Daniel. "Algorismus in GKS 1812." Master's Thesis, Háskóli Íslands, 2020. Accessed December 9, 2025. https://skemman.is/handle/1946/37284.

Sloughter, Daniel. "*Algorismus* in GKS 1812 4to." *Gripla* XXXII (2021): 73–100.

Smith, David E. *History of Mathematics*. Vol. 2. New York: Dover, 1958.

Smith, David Eugene, and Louis Charles Karpinski. *The Hindu Arabic Numerals*. Boston, MA: Ginn & Company, 1911.

Southern, R. W. *Platonism, Scholastic Method, and the School of Chartres*. Reading, England: University of Reading, 1979.

Steinunn Kristjánsdóttir. *Monastic Iceland*. London and New York: Routledge, 2023.

Steinunn Kristjánsdóttir. "The Dissolution of the Monastic Houses in Iceland." *Religions* 15, no. 7:771 (2024). Accessed December 10, 2025. https://doi.org/10.3390/rel15070771.

Svanhildur Óskarsdóttir and Vasareé Rastonis. "Fragments United: The Codicology of GKS 1812 4to." In *A World in Fragments: Studies on the Encyclopedic Manuscript GKS 1812 4to*, edited by Gunnar Harðarson, Christian Ethridge, Guðrún Nordal, and Svanhildur Óskarsdóttir, 39–62. Reykjavík: Stofnun Árna Magnússonar í Íslenskum Fræðum, 2021.

Tao (@tao), Terence. "My own personal definition of mathematics," December 5, 2024. Accessed December 10, 2025. https://mathstodon.xyz/@tao/113601397729489852.

Taylor, A. E. *A Commentary on Plato's Timaeus*. Oxford: Clarendon Press, 1928.

Thesleff, Holger. *An Introduction to the Pythagorean Writings of the Hellenistic Period*. Turku, Finland: Åbo Akademi, 1961.

Thorndike, Lynn, ed. *The Sphere of Sacrobosco and Its Commentators*. Chicago: The University of Chicago Press, 1949.

Uspensky, V. A., and A. L. Semenov. "What are the Gains of the Theory of Algorithms: Basic Developments Connected with the Concept of Algorithm and with its Appplications in Mathematics." In *Algorithms in Modern Mathematics and Computer Science*, edited by E. P. Ershov and D. E. Knuth, 100–234. Berlin and Heidelberg: Springer, 1981.

Vernet, J. "al-KHwārazmī." In *The Enclyclopaedia of Islam*, edited by C. E. Bosworth, E. van Donzel, B. Lewis, and Ch. Pellat, 4:1070–71. Leiden: E. J. Brill, 1997.

Wailly, Natalis de. *Éléments de paléographie*. Vol. 2. Paris: Imprimerie Royale, 1838.

Woepcke, M. F. *Sur l'introduction de l'artithmétique Indienne en Occident et sur deux documents importants*. Rome: Imprimerie des Sciences Mathématiques et Physiques, 1859.

Index of Personal Names

 | https://doi.org/10.1515/9781501522376-012

Index of Subjects

 | https://doi.org/10.1515/9781501522376-013